COMPUTER ARCHITECTURES FOR ROBOTICS AND AUTOMATION

COMPUTER ARCHITECTURES FOR ROBOTICS AND AUTOMATION

Edited by

James H. Graham

University of Louisville

Gordon and Breach Science Publishers

New York London Paris Montreux Tokyo

Gordon and Breach Science Publishers

Post Office Box 786
Cooper Station
New York, New York 10276
United States of America

Post Office Box 197
London WC2E 9PX
England

58, rue Lhomond
75005 Paris
France

Post Office Box 161
1820 Montreux 2
Switzerland

14–9 Okubo 3-chome
Shinjuku-ku, Tokyo 160
Japan

Library of Congress Cataloging in Publication Data

Computer architectures for robotics and automation.

Based on presentations made at 2 workshops held at the 1985 and 1986 IEEE International Conference on Robotics and Automation.
Includes bibliographies and index.
1. Robotics—Congresses. 2. Automatic control—Congresses. 3. Robot vision—Congresses. 4. Computer architecture—Congresses. I. Graham, James H.
TJ210.3.C65 1986 629.8′92 86-22783
ISBN 2-88124-154-9

CONTENTS

FOREWORD

This book addresses the important topic of providing adequate computational power for the computationally demanding real-time requirements of advanced robotic systems.

In order for robots to satisfactorily fulfill the many potential missions and applications envisioned for them, it is necessary to incorporate many of the recent advances in the fields of image processing, control and machine intelligence into the robotic system in real time. The most promising way to achieve this goal appears to be through innovative, specialized computer architectures which incorporate both parallelism and special purpose hardware in their design. This book is a collaborative effort of several active researchers who are attempting to address various aspects of this problem.

The book is divided into four sections, each covering a different aspect of the subject. The first section is an introductory overview of the field describing various high performance computing techniques, and outlining the goals for robotics and automation applications of these techniques.

The second section focuses on the question of robot control. The two papers in this section describe architectures which solve parts of the robot control problem through special hardware. The first paper describes robot motion control by a special control-oriented architecture. The second paper describes architectures for the efficient real-time computation of robot Jacobians.

The third section deals with the problems of vision and sensory processing in robotics. The first paper in the section describes a specialized pipelining architecture for robotic vision. The second paper provides an overview of vision processing algorithms and architectures. The final paper describes the use of special parallel architectures in sensory processing.

The fourth and final section includes papers dealing with two important additional areas of robotics and automation. One paper describes the question of architectures for systems of computers operating in an integrated manufacturing environment. The final paper discusses the use of a general purpose parallel architecture, the multimicroprocessor hypercube architecture, in robotics.

Much of the material in this book was presented at two workshops on Specialized Computer Architectures for Robotics and Automation held at the 1985 and 1986 IEEE International Conferences on Robotics and Automation. The editor would like to thank the contributors and participants in these workshops for their contributions: Dr. George Saridis of the IEEE Robotics and Automation council for his encouragement and support of these workshops, Mr. Harry Hayman for his logistical support and Ms. Patricia Stella and Ms. Sandra Kelty for their assistance in preparing workshop materials and sections of this book. Finally, he would like to thank the contributing authors to this book for their hard work in preparing and revising their sections. We are hopeful that this work will help point the directions for future work in these areas.

JAMES H. GRAHAM

Chapter 1

HIGH PERFORMANCE COMPUTER ARCHITECTURES FOR ROBOTICS AND AUTOMATION

James H. Graham

Department of Engineering Mathematics
and Computer Science
Speed Scientific School
University of Louisville
Louisville, Kentucky 40292

INTRODUCTION

Modern robotics offers humanity a wide array of economically and socially laudable benefits. Industrial robots are already assuming many hazardous, unpleasant or boring tasks, while simultaneously improving the productivity of factories in the United States and throughout the rest of the industrialized world. Autonomous robots can potentially handle tasks in hostile or inaccessible environments, such as, underwater, in space, or in nuclear power reactors. Household robots are presently expensive and very limited, but will eventually become more functional, less expensive and more popular. There are many other potential applications currently under study for robots, including military, prison security, and handicapped assistance.

In order for robots to satisfactorily fulfill the many potential missions and applications, it is necessary to incorporate many of the recent advances in image processing, pattern recognition, optimal and adaptive control, and artificial intelligence into *real-time* operation in the robot system. This is a worthy challenge, because many of the algorithms in these fields require hours of computation on mainframe computers, and much longer on current microcomputers, despite order of magnitude improvements which have been made in recent years in the speed and density of microelectronics. The most promising answer to this dilemma seems to be novel computer architectures which incorporate both special purpose hardware and parallelism in their designs.

The purpose of this chapter is to provide an introductory overview of the requirements posed by robotic and automation systems and the architectural alternatives that are available to meet these requirements. Hopefully this will provide an introduction to the more specialized work reported in the chapters which follow. The following section provides a brief motivation for why specialized architectures may be more appropriate than their more general counterparts. The third section presents an overview of parallel and concurrent computing architectures. The fourth section indicates the goals for such systems. An extensive, but not exhaustive, bibliography is included for those readers who wish to further pursue the topic.

MOTIVATION FOR SPECIALIZED ARCHITECTURES

Many individuals mistakenly confuse the well publicized increase in integrated circuit complexity with equivalent increases in computational throughput measured in operations per second. Figure 1 illustrates the well-known rule-of-thumb attributed to Gordon Moore, a founder of Intel, that the complexity of integrated circuitry doubles yearly. The straight line on the semilogarithmic scales indicates that exponential growth has occurred over the past twenty years, from

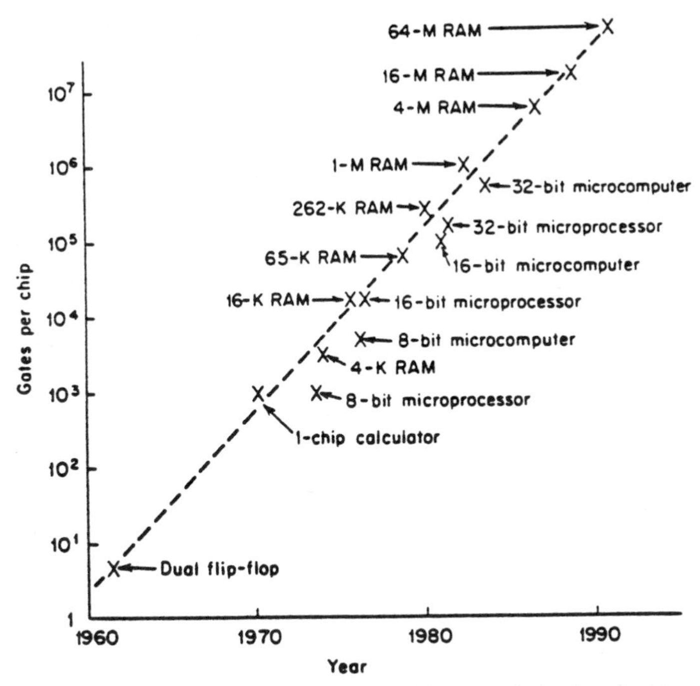

FIGURE 1 Moore's Law: The complexity of integrated circuitry doubles yearly. (FIGURE courtesy of Texas Instruments, Inc.)

fewer than 10 gates per chip in the early sixties to over 10^6 gates per chip in the 1980's.

By contrast, Figure 2 shows the increased computational power of several classes of computer systems over roughly the same time period. The scales are again semilogarithmic, but in this case the growth curves are not straight, but instead are concave downward, indicating that exponential growth is not occurring, despite the exponential growth of circuit complexity. The reason for this disparity is that communications overhead generally increases substantially with architectural complexity. Any computer system consists of several functional units which must communicate with each other in achieving an overall goal, and as circuit complexity increases, this communications activity increases steeply. In fact, a poorly designed interconnection network can very quickly lead to a point of diminishing performance returns.

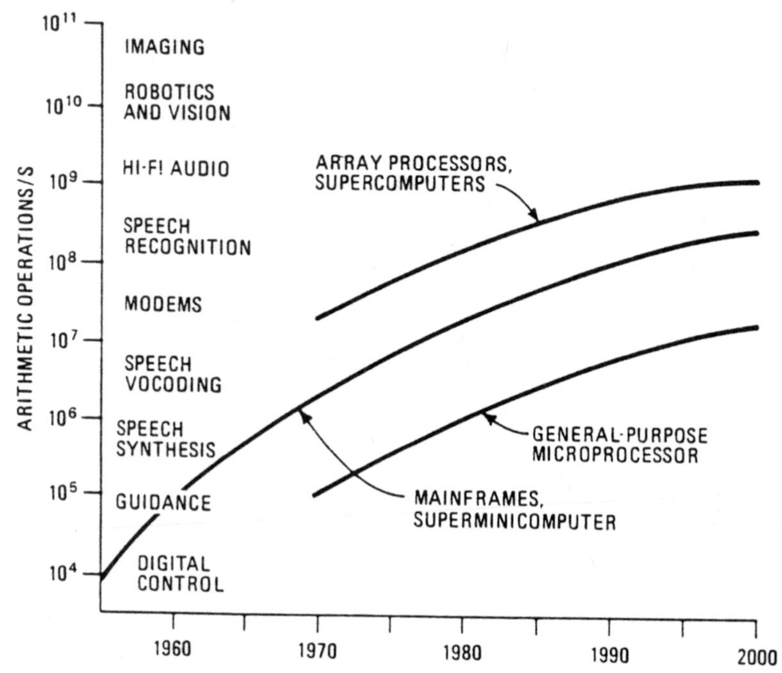

FIGURE 2 Increases in Computational power for classes of computers. (FIGURE courtesy of Texas Instruments, Inc.)

In the opinion of this author, there is one potential salvation in that the analysis so far has been based on *general-purpose* computer architectures. In fact several *special purpose*, single chip architectures have been proposed which achieve much better performance than general purpose architectures for *specific applications*. Texas Instruments, for example, currently produces a family of digital signal processing chips (the 32000 family) which achieve very high throughouts for signal processing tasks.

Although digital signal processing (DSP) chips can only solve a fraction of the computational problems of robotic and automation systems, it seems plausible that other classes of special purpose parallel architectures can be developed to complement the DSP architectures, and together they can solve a significant subset of real-time computational problems.

PARALLEL AND CONCURRENT ARCHITECTURES

As indicated in the preceding section, that best hope for overcoming the computational bottleneck of conventional processor design seems to be the development of specialized parallel architectures. This section provides an overview of some of the architectural alternatives that are available for this purpose.

There are several ways to classify parallel computing systems. One of the earliest classifications was the one proposed by Flynn which considers parallelism in instruction and data streams[6]. The four categories in the Flynn taxonomy are thus:

SISD – Single Instruction Stream
Single Data Stream

SIMD – Single Instruction Stream
Multiple Data Stream

MIMD – Multiple Instruction Stream
Multiple Data Stream

MISD – Multiple Instruction Stream
Single Data Stream

The SISD configuration corresponds to the conventional von Neumann style uniprocessor architecture, while the other three configurations represent parallel architectures. In the SIMD configuration, shown in Figure 3, the central control processor broadcasts instructions to the arithmetic processors, which are executed in parallel in lockstep fashion on separate data items. SIMD architectures are most appropriate for vector and matrix type computations. For example, in performing vector addition, each processor would fetch the appropriate components from local memory, and the component additions would be done simultaneously by all arithmetic processors. An important element of any SIMD system is the interconnection network which allows processors to exchange data[12].

In the MIMD configuration, shown in Figure 4, the central control processor is replaced by a central scheduler. In this architecture each processor fetches and executes its own instructions. It is thus

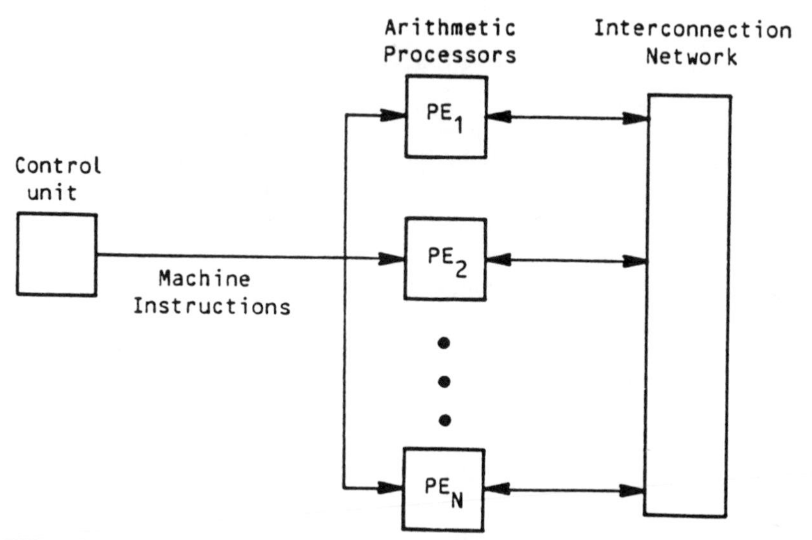

FIGURE 3 Single-instruction stream, multiple-data stream (SIMD) parallel architecture.

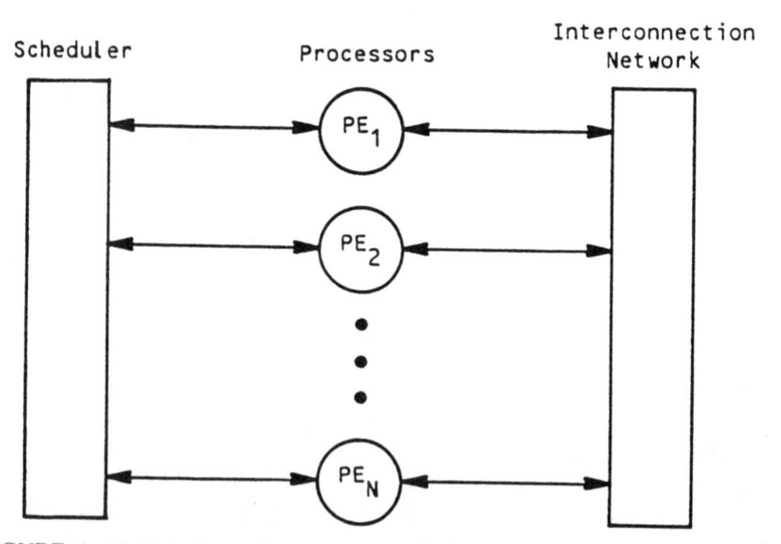

FIGURE 4 Multiple instruction stream, multiple data stream (MIMD) parallel architecture.

likely that different processors will be executing different instructions at any given instant. A MIMD system uses special hardware and software signals to coordinate the activities of individual processors. MISD architectures execute separate instruction on a common data stream, and have not been widely referenced in the literature. This form of parallelism might be appropriate for such applications as simultaneously applying a set of algorithms to a data set in a pattern matching task.

Although the Flynn taxonomy provides a good framework for an initial classification of parallel processing systems, there are many aspects of parallel and concurrent systems that are not easily classified by this taxonomy. In particular the Flynn system does not distinguish the degree of coupling of a system as measured by shared resources, such as buses, memories, clocks, etc. Systems which have a high degree of closely shared resources are commonly called tightly coupled systems, while systems with a low degree of sharing are called loosely coupled systems. SIMD architectures are typically tightly coupled because of the high synchronization required between the arithmetic processors. MIMD architectures may be either tightly or loosely coupled.

The Flynn taxonomy also fails to easily classify the type of parallelism which is found in overlap and pipelined systems. Figure 5a shows the standard execution pattern for a conventional processor; instructions are fetched, then executed, fetched, then executed, etc. A simple form of overlap processing is to allow one part of the processor to fetch the next instruction at the same time that the current instruction is being executed by the processor execution unit as shown in Figure 5b. In steady state operation, with no branches, this can increase processing speed up to 100%. Branches in the program reduce the amount of speedup, because when a branch is taken, the prefetched instruction is not valid, and the processor must wait until a fetch of the next instruction at the branch address can be completed.

A further enhancement of the fetch/execute overlap can be made by dividing the execute phase into a decode phase and an execute phase. Then, as shown in Figure 5c, additional efficiency is achieved, as three simultaneous activities take place. As instruction #1 is executed, instruction #2 is being decoded, and instruction #3 is being fetched. In this case, a branch will result in the execution unit

OVERLAPPED PROCESSING

```
(a)  | FETCH | EXECUTE | FETCH | EXECUTE | FETCH | .....

(b)  | FETCH | EXECUTE |
             | FETCH | /// | EXECUTE |
                     | FETCH | /// | EXECUTE |
                                 .
                                  .
                                   .

(c)  |FETCH |DECODE | EXECUTE |
            | FETCH | DECODE | EXECUTE |
                    | FETCH | DECODE | EXECUTE |
```

FIGURE 5 Overlapped processing. (a) Conventional sequential processing. (b) Fetch/execute overlap processing. (c) Fetch/decode/execute overlap processing.

remaining idle for two cycles as it waits for the correct next instruction to be fetched and decoded.

The idea of subdividing the execution operation leads directly to the processing speedup technique of pipelining which has been extensively used for high performance computing systems[7]. In pipelining the execution of an instruction is broken into a sequence of steps which are performed by individual hardware units in a fashion analogous to the traditional industrial assembly line.

Pipelining has been especially applicable to floating point arith-

PIPELINED PROCESSING

EX - FLOATING POINT ADD

1. COMPARE EXPONENTS

2. ALIGN MANTISSAS

3. ADD MANTISSAS

4. NORMALIZE RESULT

PIPELINED ARCHITECTURE

→ → P_1 → P_2 → P_3 → P_4 → SUM

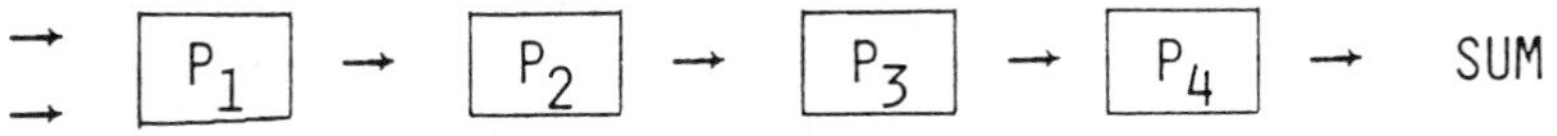

FIGURE 6 Pipelined architecture for floating point addition.

metic computations which consist of an easily partitioned set of operations. Figure 6 shows a pipelined architecture for computing the sum of two floating point operands. Processor P1 compares the exponents of the two operands, determines which is larger, and the difference between them. Processor P2 aligns the mantissas by shifting one by an amount equal to the difference of exponents. Processor P3 adds the aligned mantissos. Processor P4 then performs any post-addition normalization which is required. When the pipeline is full, one completed floating point sum will be produced on each processor cycle, as shown in Figure 7.

In theory, further performance improvement may be obtained by further subdividing the tasks and thus increasing the number of stations on the pipeline. In practice it is necessary to carefully

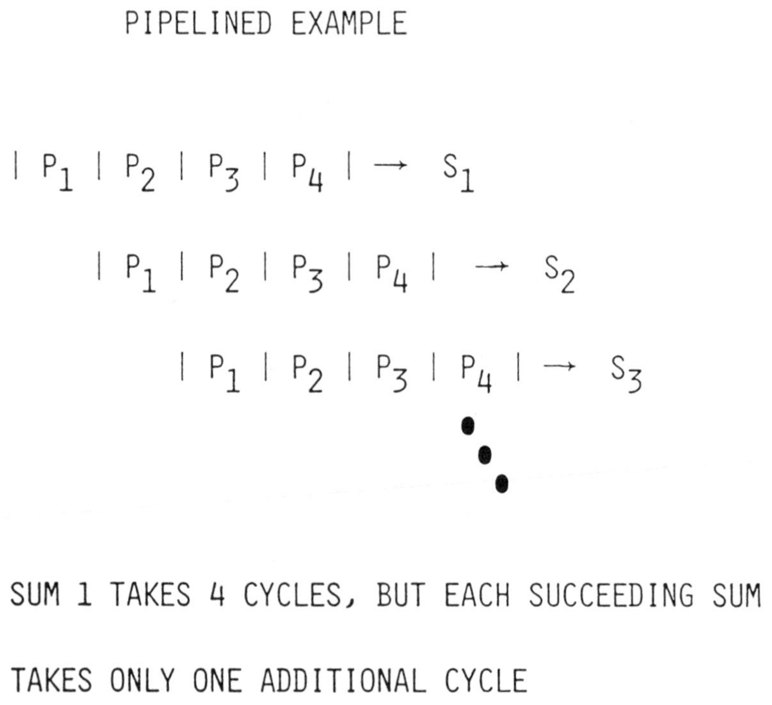

PIPELINED EXAMPLE

| P_1 | P_2 | P_3 | P_4 | → S_1

| P_1 | P_2 | P_3 | P_4 | → S_2

| P_1 | P_2 | P_3 | P_4 | → S_3

SUM 1 TAKES 4 CYCLES, BUT EACH SUCCEEDING SUM TAKES ONLY ONE ADDITIONAL CYCLE

FIGURE 7 Floating point pipeline in operation.

balance cycle times, data transfer times, and the utilization factor of the pipeline. Keeping the pipeline full a high percentage of the time is essential to achieving significant computational speedup.

A generalization of the pipelining concept is the two dimensional pipeline or systolic array[11]. In this architecture processors are interconnected in a two dimensional grid (rectangular, hexagonal, etc.) and the partial results of computations can be passed to one or more adjacent processors. Data can be entered from one or more edges of the grid and retrieved from an opposing edge.

Systolic architectures have proven advantageous for matrix and vector computations. Figure 8 shows a hexagonally connected systolic array for computing the matrix equation

$$C = A \cdot B + C.$$

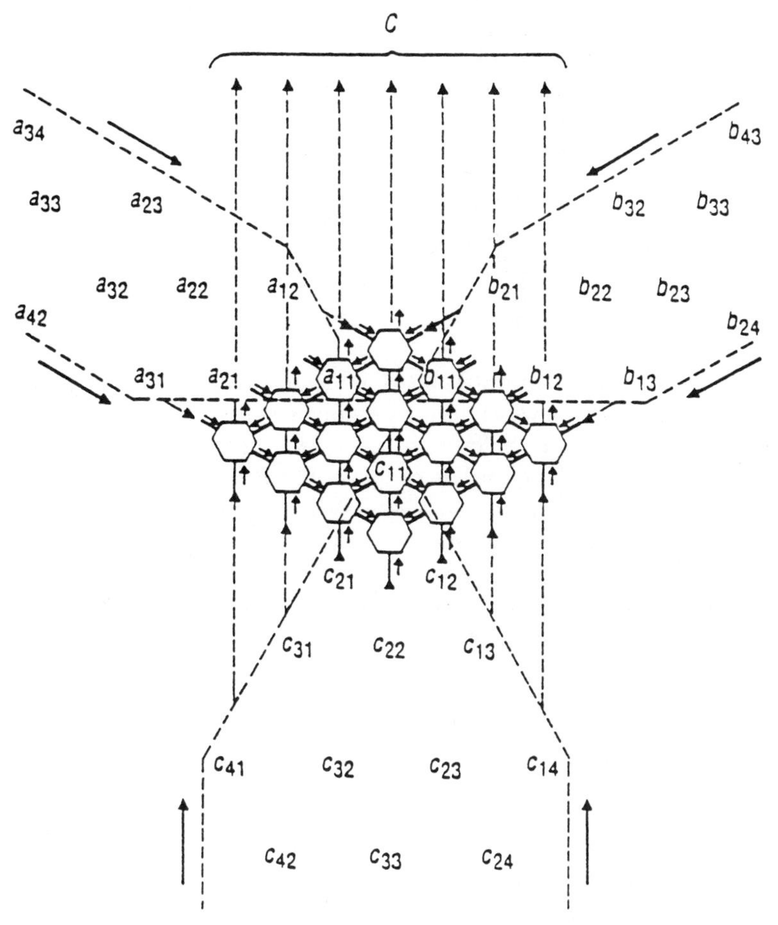

FIGURE 8 Data flow in a hexagonally-connected systolic array. (From [Haynes 82].)

It is important to recognize that the input and output data is staggered so that the proper partial products arrive at the correct processors at the proper time. The time to stagger the inputs and realign the outputs may adversely impact the theoretical performance gains.

Finally, a topic which cuts across many different forms of parallelism and concurrency is the question of choosing an appropriate

- CROSS-BAR
- RING
- HYPERCUBE
- BANYAN
- PERFECT SHUFFLE
- TREE
- PM2I
- CHORDAL RING
- CUBE CONNECTED CYCLES

Table 1 Interconnection Networks.

interconnection network. Table 1 indicates some of the many network configurations which have been investigated for various parallel systems. The major trade-off in network selection usually connectivity versus cost. In order to have efficient and time effective communication for a variety of tasks, it is necessary that a number of interconnections be established for each node, which raises the hardware cost of the network, as in the crossbar interconnection. The alternative is to have relatively few connections and to allow messages to route themselves through several intermediate nodes, as is done in a ring connection. Several intermediate networks, such as PM2I, cube connected cycles, and shuffle-exchange networks have been proposed to bridge the gap between high connectivity and low cost[12].

ROBOT APPLICATIONS FOR SPECIAL ARCHITECTURES

As the previous sections have delineated, there are many forms of parallelism and concurrency which can be applied to computational problems. Matching the appropriate parallel technology to the specific application area in a cost effective manner is an open area of engineering research and development. This section examines briefly some of the applications areas in robotics and automation where such architectures might be beneficially employed.

Figure 9 shows a simple hierarchically configured robot work-

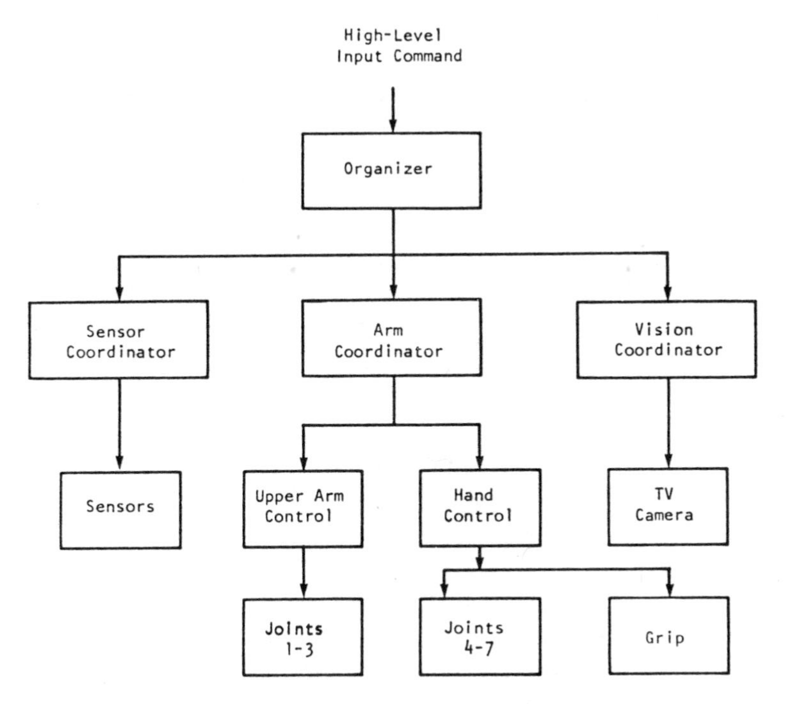

FIGURE 9 Hierarchical control scheme for a manipulator with visual and sensory feedback.

station, consisting of a robot manipulator, a machine vision system, sensory systems, and workstation coordinators. The following sections examine each of the subsystems in terms of computational needs.

Robot Control

Figure 10 illustrates the basic components of a typical industrial robot controller. In additon to the primary function of joint angle control, the robot controller is also responsible for interfacing with the human operator for set-up, programming, maintenance, etc; long-term nonvolatile program storage; interfacing to other industrial equipment; and monitoring conditions related to the safe and proper use of the robot, such as movement limits, load ranges, etc.

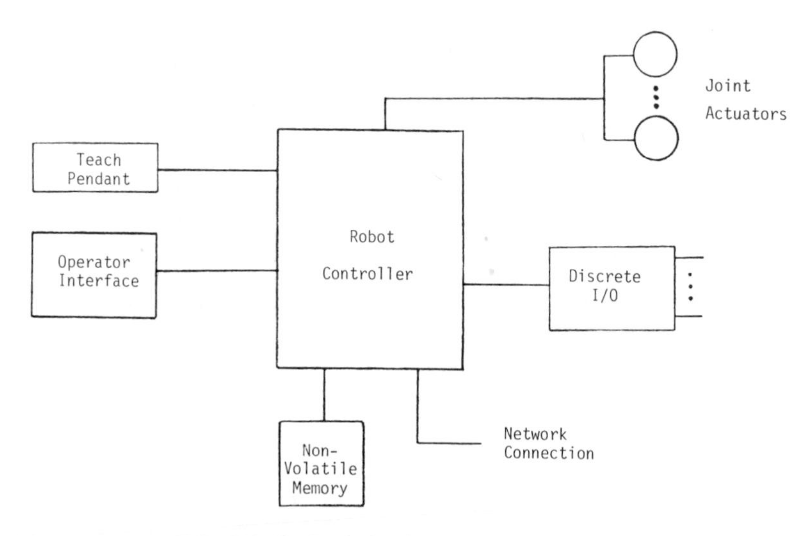

FIGURE 10 Typical robot control architecture.

From a computational standpoint, the joint control is the most demanding task. The other tasks can usually be handled by one or more conventional microprocessor systems. The computational burden of the control task is a function of the control algorithms employed, which are selected on the basis of the performance specifications for the intended robot activities. Simple pick and place material handling can usually be accomplished by use of simple proportional-integral-derivative (PID) control algorithms.

These algorithms are subject to overshoot and oscillations when pushed beyond nominal limits of speed and load. They are also subject to parameter variations. More sophisticated robot tasks in assembly applications and autonomous operations will likely require more sophisticated control algorithms. Many research programs are currently investigating adaptive, optimal, suboptimal and robust control algorithms for robots. In many cases these advanced algorithms appear suitable for vector type parallel architecture due to the vector-matrix form of their state space formulations. The topic of control is discussed in more detail in subsequent chapters.

Vision and Sensory Processing

This category of robot activity is the key to many of the advanced robotic applications which have been envisioned. The necessity for the robot to be able to interact with its environment through visual and sensory inputs generally increases proportionally with the sophistication of the proposed application.

Sensing of external forces, usually through a wrist mounted force sensing unit, is essential for active compliant assembly involving tight tolerances, force fits, etc. (Certain simple compliant operations may be accomplished with passive devices such as the remote center compliance unit.) Compliance is also necessary in a number of common activities such as turning a crank or opening a door.

Two important sensory modalities for interacting with workspace objects are tactile and proximity sensing. Tactile sensing employs force sensing over a small local area, usually the fingertips, to acquire the surface condition of external objects. Proximity sensing utilizing infrared, capacitive or ultrasound technology can be employed to safely approach external objects.

Computer processing of image data has been studied in a number of application domains with special emphasis on automation and robotics, probably because of the heavy reliance that humans place on visual inputs for our ordinary activities. Computer vision can range from simple template matching for inspection of industrial parts to complex natural scene analysis and interpretation. Recently considerable research attention has been given to systems which directly produce and utilize range information to understand the three dimensional properties of a scene.

A predominate feature of all sensory inputs is the rapid processing of large amounts of input data. Vision is especially data intensive. The input data must be rapidly and efficiently condensed to extract key features needed to make control decisions. Some of the techniques used in the digital signal processing chips seem appropriate to this task. Several chapters are devoted to architectures for sensory and visual processing.

Planning, Organization, and Coordination

The higher levels of the hierarchy shown in Figure 9 are concerned with the activities of coordination of the lower level elements so as to obtain some overall planned objective. By the hierarchical structuring principle, these higher levels operate with less detailed information and hence less precision, but with greater decision authority and (hopefully) more intelligence.

There are two architectural issues associated with these higher level functions; namely, how do we get the required data from the lower levels to the higher levels (and back), and then how do we most effectively process the data at the higher levels to obtain the control and coordination decisions? Taking the latter question first, the predominant current trend seems to be toward specialized chip architectures for executing high level planning and decision languages such as LISP and PROLOG.

The second architectural issue involves effective intercommunications between the lower direct controller elements and the higher coordinating controller elements. At the level of workstation control, a direct star connected network, with the station coordinator as the network hub, is a simple and cost effective solution to the communications problem. As more levels are added to the hierarchy, direct connections become unacceptable due to cost and reliability considerations, and other forms of computer networking, such as token passing rings and carrier-sense multiple-access networks, must be considered. This becomes a problem of choosing the best *systems architecture* to apply.

GOALS FOR SPECIAL ARCHITECTURES

In summary this chapter has attempted to outline the major forms of parallel and concurrent architectures that are either currently in use or under investigation, and how these architectures may be beneficially applied to improve the capabilities and functionality of current and future robots.

In terms of enhancing robot performance we would expect specialized architectures to support control algorithms which would make the mechanical performance of the robot faster, smoother, more accurate, and adaptable to parameter changes. At the same time, we would expect the sensory processing modules to contain other forms of specialized architectures to aid in the real-time processing of force, tactile, proximity and visual information which would permit more demanding sensory interactive activities such as compliant assembly or autonomous navigation. Finally at the level of planning, organization and coordination there is much current interest in systems architectures for coordinated supervisory control of *systems* of robots that occur in highly automated factories or in multiple-limbed walking vehicles.

Obviously much more research and development must be completed before these goals can be met in practical robots. However, it is hoped that the chapters in this book will help illuminate some directions for this effort.

REFERENCES

General References on Parallel Computing

1. IEEE Transactions on Computers — September 1980, April 1981 and November 1984 issues.
2. Journal of Parallel and Distributed Computing.
3. International Conference on Parallel Processing.
4. International Conference on Computer Design.
5. IEEE Computer Magazine — September 1981 and January 1982 issues.
6. H. S. Stone (ed), *Introduction to Computer Architecture*, Science Research Associates, 1975, Chapter 8, *Parallel Computers*.
7. K. Hwang and F. Briggs, *Computer Architecture and Parallel Processing*, McGraw-Hill, 1984.
8. D. J. Kuck, D. H. Lawrie, A. H. Sameh (eds), *High Speed Computer and Algorithm Organization*, Academic Press, 1977.
9. L. S. Haynes et al, *A Survey of Highly Parallel Computing*, Computer, Volume **15**, No. 1, January 1982, pp. 9–24.
10. D. L. Kuck, *A Survey of Parallel Machine Organization and Programming*, ACM Computing Surveys, Volume **9**, No. 1, March 1977, pp. 29–60.
11. H. T. Kung, *Why Systolic Architectures?*, Computer, Volume **15**, No. 1, January 1982, pp. 37–46.
12. H. J. Siegel, *Interconnection Networks for Large Scale Parallel Processing: Theory and Case Studies*, Lexington Books, Lexington, Mass, 1985.

Architectures for Signal Processing

13. H. T. Kung, *Special Purpose Devices for Signal and Image Processing*, Proc. SPIE, Volume **241**, July 1980, pp. 76–84.
14. H. T. Kung, L. M. Ruane, D. W. L. Yen, *A Two-Level Pipelined Systolic Array for Convolutions*, in *VLSI Systems and Computations*, Computer Science Press, October 1981, pp. 255–264.
15. H. T. Kung, *The Structure of Parallel Algorithms*, in *Advances in Computers*, Volume **19**, Academic Press, 1980, pp. 65–111.
16. S. Y. Kung et al, *Wavefront Array Processor: Language, Architecture and Applications*, IEEE Trans. Computers, Volume C-31, No. 11, November 1982, pp. 1054–1066.
17. D. J. Myers, P. A. Ivey, *STAR — A VLSI Architecture for Signal Processing*, Conf. on Advanced Research in VLSI, MIT, January 1984, pp. 179–183.
18. J. Grinberg, G. R. Nudd, R. D. Etchells, *A Cellular VLSI Architecture*, IEEE Computer, January 1984, pp. 69–81.
19. G. R. Nudd, J. G. Nash, S. S. Narayan, *An Efficient VLSI Structure for Two-Dimensional Data Processing*, International Conf. on Computer Design, 1983.
20. S. Magar et al, *An NMOS Digital Signal Processor with Multiprocessing Capability*, ISSCC, 1985.
21. R. Travassos, *Application of Systolic Array Technology to Recursive Filtering*, *Modern Signal Processing*, Prentice Hall, 1983.

Architectures for Control

22. R. Travassos, H. Kaufman, *Impact of Parallel Computers on Adaptive Flight Control*, Proc. 1980 Joint Automatic Control Conf., pp. WPI-B.
23. R. Travassos, H. Kaufman, *Parallel Initial Costate Search Algorithms for Computing Optimal Control*, Proc. 1979 Conf. on Decision and Control.
24. R. Travassos, *VLSI Implementation of Parallel Kalman Filters*, AIAA Guidance and Control Conf., August 1982.
25. T. Kailath, *Estimation and Control in the VLSI Era*, 1983 Conf. on Decision and Control, pp. 1013–1024.
26. J. H. Graham, T. F. Kadela, *New Parallel Algorithms and Architectures for Optimal State ,Estimation*, 1984 International Conf. on Parallel Processing, pp. 481– 489.
27. J. H. Graham, T. F. Kadela, *VLSI Architectures for Optimal State Estimation*, 1984 International Conf. on Computer Design, pp. 364–369.
28. J. H. Graham, G. N. Saridis, *Two Level, Multi-Microprocessor Digital Controller*, Journal of Automatic Control Theory and Applications, Volume **8**, No. 1, January 1980.
29. J. M, Jover, T. Kailath, *A Parallel Architecture for Kalman Filter Measurement Update*, 1984 IFAC World Congress.
30. V. Jaswa, C. Thomas, *An Architecture for Control*, Conf. on Advanced Research in VLSI, MIT, January 1984, pp. 21–25.
31. V. Jaswa, C. Thomas J. Pedicone, CPAC — *Concurrent Processor Architecture for Control*, IEEE Trans. Computers, Volume C-**34**, No. 2, February 1985, pp. 163–169.

Architectures for Vision and Sensing

32. K. S. Fu, *Special Computer Architectures* for Pattern Recognition and Image Processing, AFIPS National Computer Conf., June 1978, pp. 1003–1013.
33. F. A. Briggs, K. S. Fu, K. Hwang, J. H. Patel, *PM4 — A Reconfigurable Multi-processor System for Pattern Recognition and Image Processing*, Proc. of National Computer Conf., 1979, pp. 255–265.
34. H. J. Siegel et al, *Parallel Image Processing/Feature Extraction Algorithms and Architecture Emulation*, Purdue University Technical Report, TR-EE 80–57, 1980.
35. Y. W. E. Ma, R. Krishnamurti, *REPLICA — A Reconfigurable Partitionable Highly Parallel Computer Architecture for Active Multi-Sensory Perception of 3-Dimensional Objects*, International Conf. on Robotics, March 1984, pp. 176– 182.
36. M. J. B. Duff, *CLIP4: A Large Scale Integrated Circuit Array Parallel Processor*, International Joint Conf. Pattern Recognition, December 1980, pp. 503–507.
37. S. M. Goldwasser, *Computer Architecture for Grasping*, International Conf. on Robotics, March 1984, pp. 320–325.
38. M. H. Raibert, J. E. Tanner, *Design and Implementation of VLSI Tactile Sensing Computer*, International Journal of Robotics Research, Volume **1**, No. 3, Fall 1982, pp. 3–18.
39. J. S. Albus, C. R. McLean, A. J. Barbera, M. L. Fitzgerald, *An Architecture for Real-Time Sensory-Interactive Control of Robots in a Manufacturing Facility*, 4th IFAC/IFIP Symposium on Information Control Problems in Manufacturing Technology, October 1982.
40. E. Kent, M. Shneier, R. Lumia, *PIPE — Pipelined Image Processing Engine*, National Bureau of Standards, Industrial Systems Division, Gaitherburg, Maryland.

Architectures for Computer Integrated Manufacturing

41. J. A. Simpson, R. J. Hocken, J. S. Albus, *The Automated Manufacturing Research Facility of the National Bureau of Standards*, Journal of Manufacturing Systems, Volume **1**, No. 1, 1982, pp. 17–32.
42. A. J. Barbera et al, *RCS: The NBS Real-Time Control System*, Robots 8 Conf., June 1984.
43. T. L. Johnson et al, *Emulation/Simulation of a Modular Hierarchical Feedback System*, IEEE Conf. on Decision and Control, December 1982.

Chapter 2

ROBOT MOTION CONTROL WITH CPAC: A CONCURRENT PROCESSOR ARCHITECTURE FOR CONTROL

John T. Pedicone
and
Timothy L. Johnson

Control Technology Branch
General Electric Corporate Research & Development
P.O. Box 43
Schenectady, NY 12345

INTRODUCTION

The Concurrent Processor Architecture for Control, CPAC, is an experimental processor for control which is being designed and implemented at General Electric Corporate R & D. The non Von-Neumann architecture features eight concurrent on-chip functions and peak performance of approximately 5 MIPs, based on 16-bit 2's complement arithmetic. This is the result of trade-off studies which balance silicon area, performance, and general control requirements. The architecture can simultaneously support 32 16-bit sampled-data input and output channels, 128 boolean input and output channels which are polled at another rate, and a 16 bit bidirectional bus interface, eliminating bus contention and providing a much higher throughout capacity than any control processor currently available.

This brief paper provides a sketch of how CPAC could be applied to next generation robot motion control, and illustrates how the capabilities of the processor could be exploited.

A NEXT GENERATION ROBOT MOTION CONTROL ARCHITECTURE

The motion control subsystem is here defined to be the collection of real-time computational tasks, and associated hardware, which is concerned with the execution of a planned motion. With currently available hardware, this usually involves tasks on a host processor which may perform inverse kinematics in real time to convert path coordinates to joint coordinates, and/or tasks with interpolate between taught points on a path according to a commanded speed. Typically, the commanded joint coordinates are then transmitted over a controller backplane bus to each of several joint servo processors. Each of these processors performs the function of implementing a simple control law (e.g., PID), with associated limit tests, anti-windup protection, start/stop logic, and short-term joint space interpolation. Usually, the control processors don't communicate directly, so that from a multivarible control perspective the control law is decoupled. Advanced features such as compliance control are difficult to implement, because the sensor signals are forced back through the host, incurring an excessive loop delay. Thus the designer is obliged to put force/compliance control 'outside' of the position loop, rather than 'inside' the loop, which would provide much more stable and reliable performance.

One concept for a next-generation motion control architecture would be to elevate the host function, so that the host could deal with regional events in Cartesian work-space coordinates, and compute only local reference values, Jacobian element values, and path parameters for a coordinated axis control processor. These host functions require high accuracy and may be based on global features of the overall robot task. The coordinated axis control processor (e.g., CPAC), receives a local specification of the task geometry: an accurately specified reference point such as the next target point, a local Jacobian for the surrounding region, a bounding-hyperplane description of the local task geometry, and a local compliance matrix which specifies the constrained and compliant coordinate directions. The local coordinates would be specified in hybrid cartesian (or local cartesian) force-position space, following the concept introduced by Raibert and Craig [1].

The coordinated axis control processor would seek the specified target point (force/position reference) according to the commanded path velocity and compliance, using sensed joint or endpoint positions and forces/torques measured in either joint or task coordinates. The control gain matrices, which might be occasionally updated from the host, would be used along with the local Jacobian (and its inverse) in computing the local control law for all axes. Obstacle avoidance would be based on a detection of path intersections with local task surfaces based on the host's geometric model, as well as direct contact detection by binary sensors mounted either on the arm, the workpiece, or the local worksurface. Limits on joint positions and forces or torques would also be detected in real time and accomodated in an appropriate fashion (e.g., saturation or compliance of the control law). This functionality is approximately matched to the capabilities of CPAC, and represents a significant technical advance from the state of the art. It would also represent a substantial decrease in controller cost and complexity at the board level.

CPAC HARDWARE ARCHITECTURE AND CONFIGURATION FOR ROBOT CONTROL

CPAC Hardware overview

CPAC (Concurrent Processor Architecture for Control) is a special purpose microprocessor architecture designed to encompass both the discrete and continuous domains of control [2]. CPAC is divided into two concurrent processors, the CPE (Continuous Processing

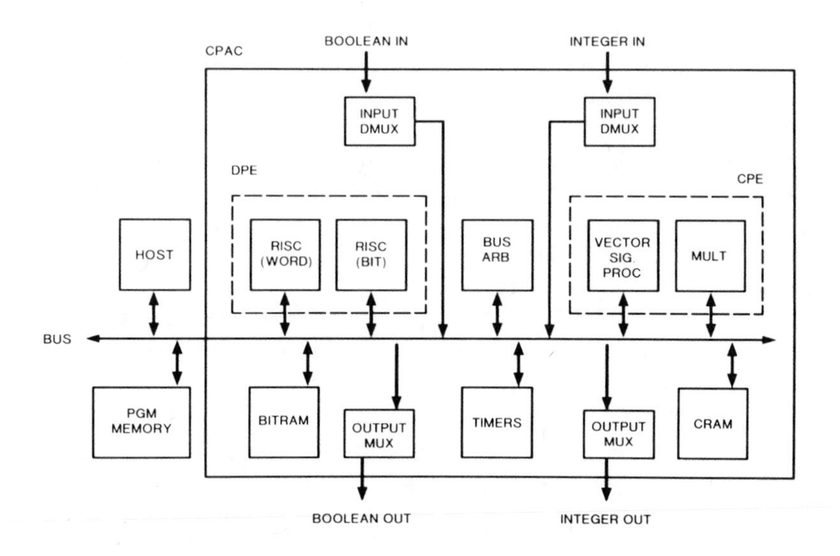

FIGURE 1 CPAC—Overview of Architecture.

Element) and the DPE (Discrete Processing Element) as shown in Figure 1. The CPE's main function is to execute control over an analog plant via linear multivariable compensator equations. The DPE primarily controls discrete inputs and outputs through the use of finite state machines and other general purpose programming techniques. The DPE'S other capability is to monitor and if necessary intervene in the operation of the CPE.

Continuous Processing Element

The CPE is composed of a hardwired vector signal processor, a multiply unit, coefficient memory (CRAM) and I/O units to transmit and receive serial data. A single CPE can accomodate a combination of 32 inputs, 32 states and/or outputs, and 128 coefficients.

In a typical application, after detection of the sample clock 16 bit integer data would be input serially through the input demultiplexor (Input DMUX) and stored in internal memory via a 16 bit bus. The calculation of the control law is concurrent with the input operation. Coefficients, inputs and states are loaded into the multiplier/accumulator (MULT) by the vector signal sequencer to compute the states and outputs of the corresponding control law. These results are stored in coefficient RAM (CRAM) and output by another concurrent process through the output multiplexor (Output MUX). CRAM is actually divided into coefficient memory, ESTATE/OSTATE, and Input RAM; the control law coefficients are stored in the coefficient memory. Inputs and references are stored in the Input RAM. ESTATE/OSTATE is used to store the present and next state vectors as well as the output vector. Control lines are provided to allow interfacing to A/D's by serial input and D/A's by serial output. An effort was also made to make the control lines general enough to enable interfacing to other CPE's and processors with minimal external circuitry.

CPE processing can be interrupted by the setting of 'breakpoint' bits by either the host or the DPE. The CPE can be instructed to enter a 'wait state' and generate an interrupt to the DPE under these conditions:

1. after the sample clock
2. input has been received
3. a single multiplication
4. a state has been calculated
5. the control law has been calculated

The CPE remains in the 'wait state' until the DPE issues an interrupt acknowledge. The purpose of these 'breakpoints' is to provide synchronization with the DPE and limit checking of input, outputs and states.

Determining boundary conditions in the state space is facilitated by 'sign registers'. After each state (or output) has been calculated, the sign of the result is stored in the 'sign registers'. This can be used to detect zero crossings. Boundary conditions can be detected by simply storing limits in Input RAM and subtracting this value from the appropriate state variable. This subtraction occurs within the calculation of the difference equations by the proper choice of coefficients. To determine if the controller is within a certain region of the state space one merely has to mask the corresponding sign bits through a DPE boolean operation. See [6].

If a different operating region is detected, coefficients can be overwritten by the DPE or by the Host. If less than half of coefficient memory is used for the control law, a new set of coefficients can be used by changing the value of Coefficient Base register (CBASE). CBASE points to the location of the first coefficient of the difference equation. By altering CBASE, a limited number of sets of coefficients can be used for different operating conditions.

Discrete Processing Element

The DPE is the logical or decision processing element of the CPE/DPE pair. Its primary function is to implement finite state machines, but it is also provided with the capability of carrying on CPE/DPE interaction, communications, and other subsidiary calculations. Unlike the CPE, the DPE is an instruction programmed machine. The DPE instruction set makes it a specialized RISC (Reduced Instruction Set Computer). While it is capable of computing any general algorithm, the instruction set reflects a desire to perform certain functions better than others. Thus the DPE is very good at boolean functions and is also very good at carrying out those functions that are typically encountered when continuous and discrete variables must interact.

The DPE part of the CPAC processing element (Figure 1) consists of a bit logic unit (BLU), a word logic unit (WLU), bit memory (BITRAM), a boolean data Input DMUX, a boolean Output MUX, a TIMER section, and a bus arbitration unit. These procsses operate concurrently to promote the highest possible throughput.

The BLU together with the WLU are responsible for the execution of DPE instructions. After instructions are fetched from program memory, an instruction is then executed by the BLU or the WLU. The BLU instructions operate on single bit operands obtained from BITRAM. A Bit Stack (a 16 bit shift register that can be shifted left or right) is used to store intermediate results during the evaluation of boolean expressions. The number of memory accesses is reduced through the use of a Bit Cache. Sixteen bit operands can be operated on without having to re-access memory. The WLU is a conventional 16 bit logic unit. It performs word operations on data contained in the DPE's 16 × 16 register file, BITRAM, CRAM or program memory.

BITRAM's function is to buffer boolean I/O, to control timers, and to be used as a 'scratch pad' for boolean operations. This section of memory is organized into 64, 16 bit words. Eight words are dedicated to boolean input and eight words are dedicated for boolean output. An additional eight words are reserved for timer control. The remaining 40 words are available for general purpose boolean variables (although BITRAM can be used for additional integer storage if desired).

The boolean input/output sections are the serial interface between the DPE and the external plant. The Input DMUX and the Output MUX can both be configured to process from 16 to 128 bits (in increments of 16 bits). Boolean inputs and outputs are stored in BITRAM and are transmitted to these sections over a 16 bit bus. Control lines are available to provide 'handshaking' between the plant and DPE. Connections between DPE's is also possible through these control lines. Boolean input occurs on the rising edge of the DPE Sync clock. The input process is enabled though a control bit. When the data has been received and stored in BITRAM, an interrupt is generated causing a jump to the associated interrupt routine. The output process is initiated when a programmable control bit is set.

The TIMER section consists of logic to time boolean or continuous (in the CPE) events. A 64 word RAM is used to store the times to be counted down. The number of timers to be used is designated by loading the control register TNUM. The rest of this memory can be used as general purpose RAM. A timer must first be loaded with the number of DPE Sync clocks to be counted. One must set a bit in BITRAM to enable a particular timer. When a DPE Sync is received each enabled timer is decremented. If any timer counts down to zero, an interrupt is generated and the number of the expired timer is stored in a register.

The Bus Arbitrator (BA) determines which concurrent process is allowed access to the CPE, DPE and the external busses. Except for the WLU, processes only access resources (memory or registers) that are connected to their corresponding element's bus (e.g. the TIMER section only accesses memory in the DPE and does not access CPE or program memory). Each process must request the bus(ses) required and be granted access by the BA. In order to speed up processing the CPE's vector signal processor can grab the CPE bus if it is not currently being used and inform the BA after the fact. The BLU and the WLU can also grab the DPE bus in a similar manner.

CPAC Hardware Architecture for Robot Control

A CPAC configuration for robot motion control is shown in Figure 2. Position and force/torque measurements for individual axes are multiplexed onto a single high-speed (5 Mbit/sec.) serial channel, providing perhaps 6–12 CPE inputs; the CPE supplies channel address generation to minimize interface hardware. The CPE outputs, similarly, are demultiplexed through double-buffered D/A converters, providing current or velocity signals to the servomotors at each joint. In addition, reference signals for sensors or external devices may be supplied. The CPE sample clock governs the common servo update rate for all axes, typically about 2 msec. The CPE provides necessary control signals to support true synchronous sampling at the input and output.

The DPE inputs might consist of overcurrent or out-of-range binary detector or limit switch signals for each joint or for the transducer, in addition to logic signals indicating end-effector integrity, proximity detection, or surface contact. DPE outputs might

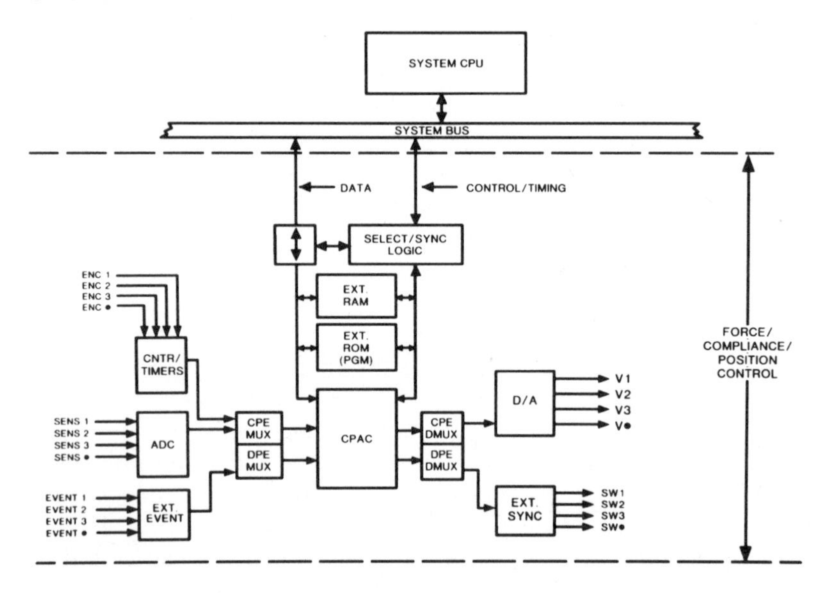

FIGURE 2 Robot Motion Control Configuration.

include activating signals for equipment, such as assembly equipment in use with the robot, alarm activation signals, end-effector release or activation signals, and brake release or activation signals. These signals are also serially multiplexed and synchronously polled, rather than interrupt- driven. The rate for DPE polling can be faster than the rate for CPE sampling, depending on the timing requirements of the associated hardware; for instance, a regular polling cycle time of 125 microseconds, or 16 times the CPE sampling rate, would be comparable with the interrupt service latencies supported by current commercial microprocessors — however, current commercial processors cannot simultaneously respond to events on such a large number of channels. The DPE provides signals for synchronous sampling of binary I/O channels (although this is not required in some applications); however, it does not provide address generation because this is usually unnecessary. The available control signals of the DPE can be used with external interface circuitry to generate addresses in the few instances where this is required.

CPAC SOFTWARE ARCHITECTURE FOR ROBOT CONTROL

CPAC Software Overview

Due to the very different nature of the CPE and DPE processors, each one is programmed in a different fashion. The CPE is 'programmed' by simply loading the coefficients of the difference equations into coefficient memory. The DPE has a reduced instruction assembly language which includes only essential instructions needed for digital controllers.

CPE Coefficients

The coefficients required for the linear difference equation are composed of two 16 bit words of coefficient memory. Twenty four of these bits contain a fixed-point number. The number of binary fraction bits is determined by the value stored in the Binary Fraction Bits register (BFB). The remaining 8 'tag bits' are used to associate a coefficient with a particular state variable or input and flag the last coefficient of that variable. These 'tag bits' result in the user not having to store zero coefficients. During the calculation of a state variable, 40 bits are accumulated. If overflow occurs, the result is set to full scale positive or negative depending on the sign of the result. After the end of a state calculation, the result is rounded to 24 bits with BFB fraction bits.

Bit Manipulation Instruction Set

The bit-manipulating instructions perform Boolean functions on discrete operands. The instruction set contains a test and branch instruction — JMPL, which allows implementation of quick tests for binary decision diagram implementation of Boolean equations. The other bit instructions perform boolean operations using the BITCACHE, the BITSTACK, and BITRAM. (See Table 1.)

ANDB	and bit
ORB	or bit
NOTB	not bit
JMPL	jump
PUSH	push bit
POP	pop bit
SETB	set bit
CLRB	clear bit

Table 1 Bit Instruction Mnemonics

Word Manipulation Instruction Set

All word-manipulating instructions are 32-bit (double-word) in length and are used for operating on all registers (DPE and CPE) and word memory locations in the CPE, DPE and program memory. (See Table 2.) Most instructions like MOV and ADD are allowed to access and operate on BITRAM using word access. The CMP (compare) instruction can also directly modify individual bits in BITRAM.

ADD, ADDC	add, add with carry
SUB, SUBB	subtract, subtract with borrow
ANDW,ORW,NOTW	and/or/not word
SHL	shift left
SHR	shift right
MOV	move (register/memory)
CALL	call
JMP	jump
RET	return
LDR	load register
STR	store register
CMP	compare
STIM	set timer

Table 2 Word Instruction Mnemonics

Example Program

The following example is a DPE assembly program (Table 3) for monitoring a single PID loop. The PID loop calculations occur concurrently in the CPE. The PID equation has been translated into a state space representation with coefficients stored in CPE memory. The program also monitors boolean inputs from a simple motor.

Interrupts are sent from the CPE after each state is calculated. These events result in a call to the 'subiteration' interrupt routine (INT_SUB in the example). In INT_SUB the process input is checked to see if it has exceeded a predetermined limit. An alarm flag (HI_ALARM) is set if the input is out of range.

Another interrupt occurs when an 'iteration' has been completed by the CPE (i.e. the control law has been calculated). In the example, the interrupt routine INT_IT is executed. The controller output is compared with a maximum limit. If this limit has been exeeded, the controller output is set to this maximum.

CPAC Motion Control Software

A block diagram for CPAC DPE motion control software is shown in Figure 3. This program is executed on every DPE polling clock cycle. For simplicity, assume that the DPE polling clock is synchronized with the CPE sample clock, but is at some multiple of its rate. Upon entry, the program tests for new communciations from the host, and tests or resets timer values. The timer values are used to count elapsed DPE cycles since the last CPE sample clock, and to keep track of elapsed time (in DPE polling cycles) since recent events have occurred. Elapsed timers may indicate that certain priority background tasks must be run during this polling cycle. Next, the boolean inputs are tested; routine outputs are generated, and software flags activating certain functions are stored in the bitram or in DPE registers. During this interval, typically, the CPE may be computing the control law, projected distances to obstacles, and tracking errors, in a completely concurrent mode; during this process, the DPE may be interrupted for brief periods to test limits on intermediate values of input, state, or output variables, and to impose saturation or to set flags if necessary. After the DPE has completed its time-critical functions, it tests to determine whether the CPE computation cycle is complete. If the CPE cycle is complete, additional tests may be performed to test and/or saturate output values before they are sent off-chip, to test for anticipated collision conditions, and to test for attainment of the commanded reference position, force, and/or torque. If the host command has been satisfied, then new gains or reference values are sought; these pertain to the next commanded end-effector condition. Finally, various background tasks may be activated, such as status logging, calibration, error detection, etc.

Table 3 Example Program for PID Loop and Motor Logic

```
            SEGM  DISEG
HOT__FLG    DS    1           ;RESERVE SPACE FOR MOTOR OVERHEATED FLAG
START       DS    1           ;1 = Start motor, 0 = stop motor
            END
            SEGM  DOSEG
HI__ALARM   DS    1           ;HI Alarm
RUN__FLG    DS    1           ;1 = Start motor, 0 = stop motor
            END
            SEGM  FLAGSEG
F__IT       DS    1           ;Flag iteration calculated
TEMP        DS    1           ;Temporary flag
            END
            SEGM  CONSTSEG    ;CONSTANTS AND ADDRESSES
HILIM       EQU   58          ;ADDR OF HILIM
ESTATE      EQU   513         ;ADDR OF NEXT STATE VALUES + 1
OSTATE      EQU   545         ;ADDR OF NEXT STATE VALUES + 1
IRAM        EQU   577         ;ADDR OF INPUT RAM + 1 (REF PID 1)
WPROC       EQU   1009        ;ADDR OF CPE BKPT FLAGS
OREG        EQU   1016        ;ADDR OF CPE OUTPUT SHIFT REGISTER
BIT__ORAM   EQU   8           ;ADDR OF DPE OUTPUT FLAGS
BIT__FRAM   EQU   32          ;ADDR OF DPE GENERAL FLAGS
;CONSTANTS
CONT        EQU   2           ;MASKS FOR SETTING BITS IN DCTRL
FIO         EQU   4
INTE        EQU   8
SCYINTFIO   EQU   13          ;START CYCLE, ENABLE INTR., START DISCRETE I/O
CLRCONT     EQU   OFDH
INTEFIO     EQU   12
            END
```

```
;Main program — Calculate 1 PID loop and monitor 1 motor
            ORG   1024
            JMP   INIT                           ;Restart Vector
            JMP   NULL_INT                       ;Interrupt Vectors
            JMP   NULL_INT
            JMP   NULL_INT
            JMP   INT_SUB                        ;Subiteration Vector
            JMP   INT_IT                         ;Iteration Vector
INIT        STR   0, ESTATE                      ;CLEAR PRESENT STATE VECTOR
            STR   0, ESTATE + 1
            LDR   VREF1, RO                      ;STORE SETPOINTS FOR PID
            STR   RO, IRAM
            LDR   VHILIM, RO                     ;STORE PROC. VAR. HIGH LIMIT
            STR   RO, HILIM
            LDR   VYLIM, RO                      ;STORE CONTROLLER OUTPUT LIMIT
            STR   RO, YLIM
            LDR   OOH, DIOW                      ;SETUP FOR 1 OUTPUT WORD, 1 INPUT
            LDR   SCYINTFIO, DCTRL               ;ENABLE CPE, INTERRUPTS, START DISCRETE I/O
            LDR   BIT_FRAM, BRPTR                ;SET BITRAM POINTER TO DPE FLAGS
MAIN        CLRB  F_IT                           ;CLEAR FULL ITERATION FLAG
            LDR   18H, R8                        ;enable SUB and IT breakpoints
            STR   R8, WPROC
            LDR   OSTATE, RO                     ;Pointer to next state error
            LDR   COND (ODD = T), ESTATE, RO     ;SET POINTER TO ESTATE error term
            LDR   IRAM, R2                       ;POINTER TO I-RAM (REF FOR PID 1)
            LDR   0, R7                          ;CLEAR SUBITERTION COUNT
            CLRB  F_OL                           ;CLEAR CONTROLLER OUTPUT LIMIT FLAG
            LDR   [HILIM], R3                    ;LOAD PTR TO HIGH LIMIT VALUE
WAIT_FIO    ANDW  SCC, DCTRL, FIO, R8            ;R8 → 0?
            JMP   COND (Z = F), WAIT_FIO         ;IF FLUSH I/O NOT COMPLETE, WAIT
            NOTB  TOS, HOT_FLG                   ;TEST MOTOR, CHECK FOR HOT_FLG
            ANDB  TOS, START                     ;CHECK FOR START
            POP   RUN_FLG                        ;STORE RUN_FLG
```

```
;
SPIN2       JMP    COND (F_IT = F), SPIN2            ;Wait for end of iteration
            ORW    DCTRL, INTEFIO, DCTRL             ;ENABLE INT., START DPE SERIAL I/O
            JMP    MAIN
;
INT_SUB     ADD    R7, 1, R7                         ;INCREMENT SUBITERATION COUNT
            CMP    R7, GT, 1, T, TEMP, BRPTR         ;CHECK FOR FIRST SUB.
            JMP    COND (TEMP = F), HILIMIT
            LDR    10H, R8                           ;DISABLE SUBITERATION BRKPT
            STR    R8, WPROC
            JMP    ENB_INT
HILIMIT     MOV    [R2], 1, R4                       ;GET 2nd I-RAM VALUE (PROC. VAR.)
            LDR    BIT_ORAM, BRPTR                   ;SET BITRAM POINTER TO OUTPUT FLAGS
            CMP    R4, GT, R3, T, HI_ALARM, BRPTR    ;HI LIMIT CHECK
            LDR    BIT_FRAM, BRPTR                   ;SET BITRAM POINTER TO DPE FLAGS
ENB_INT     ORW    DCTRL, CONT, DCTRL                ;ALLOW CPE TO CONTINUE
            ANDW   DCTRL, CLRCONT, DCTRL             ;LOWER CONT
            ORW    DCTRL, INTE, DCTRL                ;ENABLE INTERRUPTS
END_SUB     RET
;
INT_IT      SETB   F_IT                              ;END OF CPE ITERATION
            ADD    R1, 1, R2                         ;PTR TO OUTPUT
            MOV    [R2], 0, R6                       ;GET OUTPUT
            LDR    [YLIM], R8                        ;GET OUTPUT LIMIT
            CMP    R6, GT, R8, T, TEMP, BRPTR        ;CHECK IF OUT LIMIT EXCEEDED
            JMP    COND (TEMP = F), DONE             ;IF LIMIT NOT EXCEEDED, DONE
            MOV    R8, 0, [R2]                       ;SET OUTPUT TO OUTPUT LIMIT
            STR    R8, OREG                          ;SET CPE OUTPUT SHIFT REG. TO LIMIT
DONE        ORW    DCTRL, CONT, DCTRL                ;ALLOW CPE TO CONTINUE
            ANDW   DCTRL, CLRCONT, DCTRL             ;LOWER CONT
            ORW    DCTRL, INTE, DCTRL                ;ENABLE INTERRUPTS
            RET
NULL_INT    RET    ;NULL INTERRUPT
```

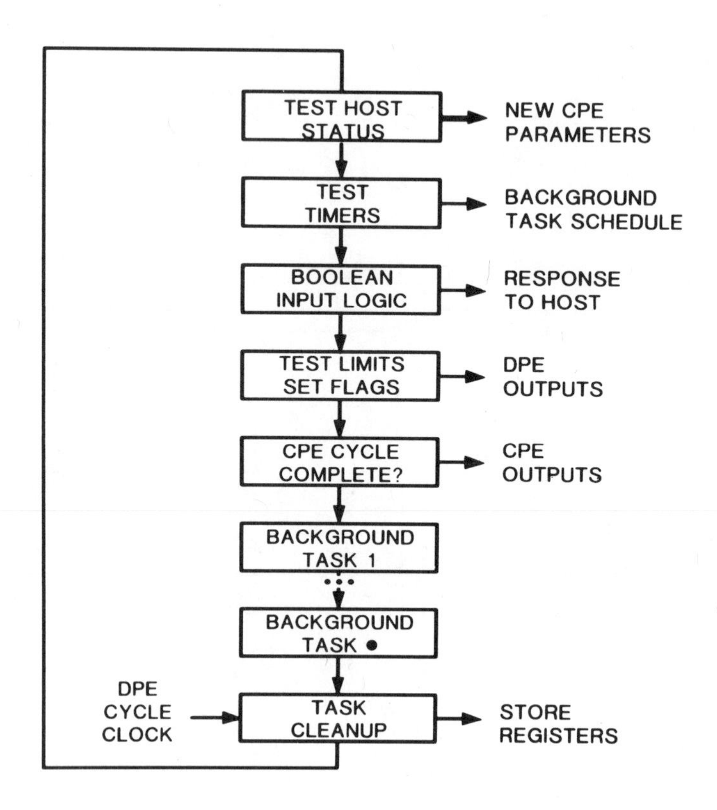

FIGURE 3 CPAC Motion Control Software Block Diagram

CONCLUSION

By sketching the architecture of a next-generation robot controller, the potential of CPAC to dramatically extend capabilities beyond those of current robot motion control subsystems has been demonstrated. Furthermore, an outline of CPAC capabilities has been provided, both in terms of hardware and software. Sample CPAC/DPE chips are expected to become available in 1987.

REFERENCES

1. Raibert, M.H. and Craig, J.J., *Hybrid Position/Force Control of Manipulators*, *J. Dynamic System, Measurement, Control* **102** (June, 1981), pp. 126–133.
2. Jaswa, V. et al., CPAC–Concurrent Processor Architecture for Control, *IEEE Computer Transactions* (February, 1985) pp. 163–169.
3. Brady, M., et al, (ed.) *Robot Motion Control* The MIT Press, Cambridge, MA, 1982.
4. *Proceedings 22nd IEEE Conference on Decision and Control*, San Antonio, TX, December 1983, pp. 1464–1466.
5. Johnson, T.L., *Synchronous Switched Linear Systems*, *Proceedings 24th IEEE Conference on Decision and Control*, December 1985, pp. 1699–1701.

Chapter 3

SYSTOLIC ARCHITECTURES FOR COMPUTATION OF THE JACOBIAN FOR ROBOT MANIPULATORS

David E. Orin, Karl W. Olson,
Hung-Hsiang Chao

Department of Electrical Engineering
The Ohio State University
Columbus, OH 43210

INTRODUCTION

Many advanced control schemes for robot manipulators are difficult to implement in real time because they involve considerable computation. Often the difficulty is based on the need to implement the kinematic or dynamic equations of motion for the manipulator. Certain of these relationships are found in a number of the approaches to advanced control and in that sense may be said to be generic computations. One of the most important of these for robot manipulators is the Jacobian. This chapter will address issues in implementing special-purpose, systolic architectures for the Jacobian.

The Jacobian relates joint rates to end-effector rates and may be used if control is based on Resolved Rate [1] or if position feedback is closed at the end-effector as in Jacobian Control [2]. It is also applicable for control of redundant manipulators, and approaches to control based on the Jacobian have been proposed in [3–6]. Furthermore, the transpose of the Jacobian relates end-effector forces to joint torques and may be used when force control is needed as in Resolved Motion Force Control [7] or in Hybrid Control [8]. While the use of the Jacobian often eliminates the need for an inverse kinematics procedure for position, a number of numerical approaches to solving inverse kinematics are based on repetitive use of the Jacobian [9,10].

Several efficient algorithms for computing the Jacobian have been detailed in [11]. However these are serial algorithms, and implementation on a uniprocessor may be difficult to achieve in real time at a reasonable cost-performance ratio. Pipeline and parallel algorithms which are appropriate for implementation on a systolic array of processors have been developed and will be given in the second section of the chapter. Following that a number of systolic architectures for the Jacobian will be given and evaluated in terms of compute time, initiation rate, and CPU utilization. In each case, the values for these criteria will be expressed as a function of the total number of degrees of freedom of the manipulator, N.

In order to provide the basic processing element for robotics computations implemented through use of systolic architectures, the design of a Robotics Processor chip has been developed at The Ohio State University [12]. It is based on custom VLSI concepts and includes a 32-bit floating point adder and 32-bit floating point multiplier. It is a Reduced Instruction Set Computer (RISC) and has a simple but efficient input/output structure which is appropriate for systolic arrays. Further details of the design are given in the fourth section of the chapter.

In order to properly evaluate the speed and efficiency of an algorithm implemented with a systolic array, it is necessary to account for the communications overhead. In the fifth section of the chapter, a more detailed evaluation of the systolic architectures for the Jacobian is made. This is based on the specifics of the Robotics Processor and includes an assessment of the overhead associated with communications between processors. Again, general formulas for the speed of computation are given and expressed as a function of the number of degrees of freedom, N. Also, the specific values for a seven degree-of-freedom manipulator are given.

Notation

The basic parameters used to describe the kinematics of a manipulator, as first presented by Denavit and Hartenberg [13] and extended and given in detail in Paul [14] are used throughout. In particular, a coordinate system is attached to each link of the manipulator with the *z*-axis directed along the joint axis. The links are numbered from 0 at the base to N at the end-effector. A separate coordinate system, labeled with the subscript N + 1 (or E for end-effector) is also fixed to the end-effector at any desired point.

The relative position of joint i, q_i, is with respect to the *z*-axis of the previous link, $\mathbf{z}_{i-1}$. That is, links $i-1$ and i are connected at joint i.

Four parameters are used to describe each successive joint and link pair — the joint angle (θ_i) and offset distance (d_i) as well as the link length (a_i) and twist (α_i) (see Fig. 1). From these parameters, the 4×4 homogeneous transformation, ${}^{i-1}T_i$, which relates positions in coordinate system i to those in coordinate system $i - 1$, may

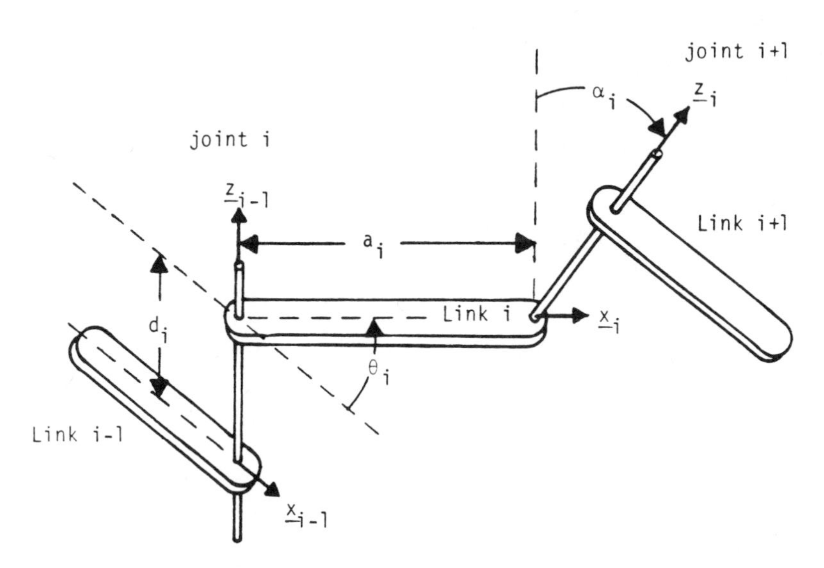

FIGURE 1 Link Parameters Associated With Link i.

be computed. For either a revolute or a sliding joint the result is as follows [14]:

$$
{}^{i-1}T_i = \left[\begin{array}{ccc|c} c\theta_i & -s\theta_i c\alpha_i & s\theta_i s\alpha_i & a_i c\theta_i \\ s\theta_i & c\theta_i c\alpha_i & -c\theta_i s\alpha_i & a_i s\theta_i \\ 0 & s\alpha_i & c\alpha_i & d_i \\ \hline 0 & 0 & 0 & 1 \end{array}\right] \quad (1)
$$

where $c\theta_i$ and $s\theta_i$ indicate the cosine and sine, respectively.

The ${}^{i-1}T_i$ matrix gives both position and orientation changes between successive coordinate systems. The top-left 3 × 3 part of the

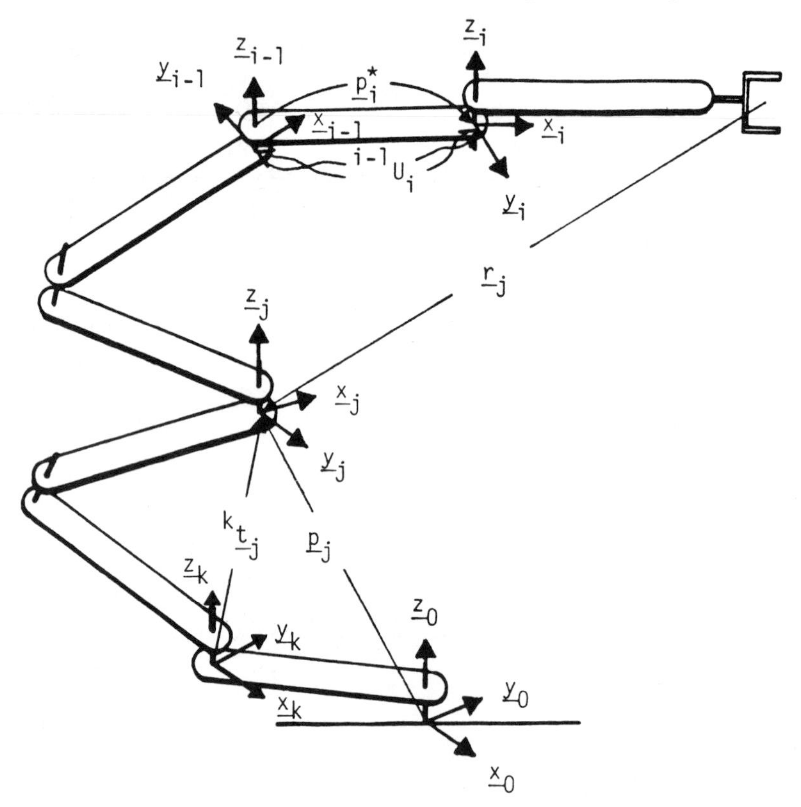

FIGURE 2 Depiction of Vectors p_j, r_j, p_i^*, ${}^kt_j$ and Transform ${}^{i-1}U_i$ for a Manipulator.

matrix gives all orientation information and will be denoted as ${}^{i-1}U_i$. The first three elements of the right-hand column give all the relative position information and will be denoted as ${}^{i-1}\boldsymbol{p}_i^*$.

The matrix ${}^kT_j$ will be used to describe the general transformation between coordinate systems associated with links j and k and may be obtained through multiplication of the intermediate transformation matrices. Following the previous notation given for orientation, the top-left 3 × 3 part of this matrix is given as ${}^kU_j$. The first three elements of the right-hand column give the position vector from the origin of coordinate system k to the origin of coordinate system j with the components given in terms of coordinate system k. It is designated as ${}^k\boldsymbol{t}_j$. If k = N + 1, then the vector is from the end-effector to link j and this is designated as $\boldsymbol{r}_j$. Similarly, $\boldsymbol{p}_j$ is from the base (link 0) to link j. These are ali shown in Fig. 2.

JACOBIAN ALGORITHMS

The Jacobian, J, may be used to transform joint rates to end-effector rates:

$$\begin{bmatrix} {}^{N+1}\boldsymbol{\omega} \\ {}^{N+1}\boldsymbol{v} \end{bmatrix} = {}^{N+1}J\ [\dot{\boldsymbol{q}}], \tag{2}$$

where

$$J = \begin{bmatrix} \gamma_1^x & \gamma_2^x & \cdot & \cdot & \cdot & \gamma_N^x \\ \gamma_1^y & \gamma_2^y & \cdot & & & \gamma_N^y \\ \gamma_1^z & \cdot & & \cdot & & \gamma_N^z \\ \beta_1^x & \cdot & & & \cdot & \beta_N^x \\ \beta_1^y & \cdot & & & & \beta_N^y \\ \beta_1^z & & & & & \beta_N^z \end{bmatrix} \tag{3}$$

and $\boldsymbol{\omega}$ is the rotational velocity vector of the end-effector, $\boldsymbol{v}$ is the translational velocity vector of the end-effector, and $\dot{\boldsymbol{q}}$ is the vector of joint rates. The components of the end-effector velocity and the Jacobian are expressed in the end-effector coordinate system

(N + 1), as indicated by the leading superscript. While the components of the end-effector velocity and Jacobian may be expressed in other coordinate systems (especially the base), the form given is especially useful in manipulator control [7] and will be used throughout this chapter. It may be noted that other forms, if needed, involve similar computations.

Serial Algorithm

The method for computing the Jacobian as proposed by Orin and Schrader [11] is especially efficient when the components of the Jacobian are to be expressed in the end-effector coordinate system. The method involves a recursive set of equations for computing the position and orientation of the $(i-1)^{st}$ link from the same values for the ith link. The equations are given as follows for the case of either a sliding or revolute joint [11]:

$$^{N+1}U_{N+1} = I; \tag{4}$$

$$^{N+1}U_{i-1} = {}^{N+1}U_i \; {}^{i-1}U_i^T, \quad i = (N+1), \ldots, 2, 1; \tag{5}$$

$$^{N+1}\boldsymbol{\gamma}_i = {}^{N+1}U_{i-1} \begin{bmatrix} 0 \\ 0 \\ 1 \end{bmatrix}, \quad i = 1, 2, \ldots, N\} \text{ revolute joint}; \tag{6}$$

$$^{N+1}\boldsymbol{\gamma}_i = \boldsymbol{0}, \quad i = 1, 2, \ldots, N\} \text{ sliding joint}; \tag{7}$$

$$^{N+1}\boldsymbol{r}_{N+1} = \boldsymbol{0}; \tag{8}$$

$$^{N+1}\boldsymbol{r}_{i-1} = {}^{N+1}\boldsymbol{r}_i - {}^{N+1}U_i \, {}^{i}\boldsymbol{p}_i^*, \quad i = (N+1), \ldots, 2, 1; \tag{9}$$

$$^{N+1}\boldsymbol{\beta}_i = {}^{N+1}\boldsymbol{\gamma}_i \times (-{}^{N+1}\boldsymbol{r}_{i-1}), \tag{10}$$

$$i = 1, 2, \ldots, N\} \text{ revolute joint;}$$

$$^{N+1}\boldsymbol{\beta}_i = {}^{N+1}U_{i-1} \begin{bmatrix} 0 \\ 0 \\ 1 \end{bmatrix}, \tag{11}$$

$$i = 1, 2, \ldots, N\} \text{ sliding joint.}$$

Note that the components of all vectors are transformed to end-effector coordinates (N+1) so as to give the components of the Jacobian in end-effector coordinates.

Pipeline Algorithm

Noting Eqs. (4) through (11), the position and orientation of the $(i-1)^{st}$ link may be computed as soon as the position and orientation of the ith link are available. Also, the components of the ith column of the Jacobian matrix may be computed as soon as the position and orientation of the $(i-1)^{st}$ link are available. This leads to the development of a pipeline algorithm for the Jacobian. Eqs. (4) through (11) may be rewritten to explicitly indicate the pipelined flow of the computations.

For $i = N, N-1, \ldots, 1$:

$$^{N+1}U_{i-1} = {}^{N+1}U_i \; {}^{i-1}U_i^T; \tag{12}$$

$$^{N+1}\boldsymbol{r}_{i-1} = {}^{N+1}\boldsymbol{r}_i - {}^{N+1}U_i \; {}^{i}\boldsymbol{p}_i^*; \tag{13}$$

$$^{N+1}\boldsymbol{\gamma}_i = {}^{N+1}U_{i-1} \begin{bmatrix} 0 \\ 0 \\ 1 \end{bmatrix}, \quad \text{revolute joint;} \tag{14}$$

$$^{N+1}\boldsymbol{\gamma}_i = \boldsymbol{0}, \quad \text{sliding joint;} \tag{15}$$

$$^{N+1}\boldsymbol{\beta}_i = {}^{N+1}\boldsymbol{\gamma}_i \times (-{}^{N+1}\boldsymbol{r}_{i-1}), \quad \text{revolute joint;} \tag{16}$$

$$ {}^{N+1}\boldsymbol{\beta}_i = {}^{N+1}U_{i-1} \begin{bmatrix} 0 \\ 0 \\ 1 \end{bmatrix}, \qquad \text{sliding joint.} \tag{17} $$

Note that it is assumed that the position and orientation of the end-effector is fixed to the last link of the manipulator (link N) so that ${}^{N}U_{N+1}$ and ${}^{N}\boldsymbol{p}^*_N$ are known and constant. Also, the index i (stage number in the pipeline) shows the flow of the pipeline as it goes from N to 1.

A block diagram explicity showing the flow of data in the pipeline algorithm is given in Fig. 3. The ith stage of the pipeline samples the angle (or linear position for a sliding joint) for the ith joint and from a knowledge of the position/orientation for the ith link, computes these values for the $(i-1)^{st}$ link. The elements for the ith column of the Jacobian matrix are then computed. Note that the communication of data between each of the stages is local, and this facilitates implementation with systolic arrays.

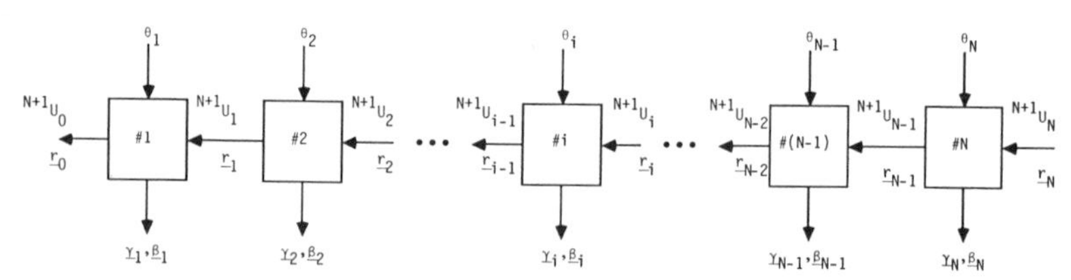

FIGURE 3 Block Diagram Showing the Flow of Data in the Pipeline Algorithm.

Parallel Algorithm

Fig. 3 for the pipeline algorithm indicates that the delay from input at stage #N to output at stage #1 will increase linearly with the number of degrees of freedom, N. A parallel algorithm will be developed in this section which will reduce the delay to the order of $\log_2 N$.

The parallel algorithm is based on a 'divide and conquer' strategy. During the first step of the algorithm, the position and orientation across pairs of links of the manipulator are computed. In the second step of the algorithm, the position and orientation across sets of four links are computed. This process is continued until at the last step the position and orientation of the base (with respect to the end-effector) are computed. The total number of steps or levels in the computational process, ℓ_T, is given as follows:

$$\ell_T = \log_2 (N+1). \tag{18}$$

Note that the end-effector is counted separately as a link.

Level	Number of Sets (m_T)	Set Width (w)
1	(N+1)/2	2
2	(N+1)/4	4
3	(N+1)/8	8
⋮	⋮	⋮

Table 1 The Number of Sets (m_T) and Number of Links per Set (Set Width, w) for Each Level of Computation in the Parallel Algorithm.

The total number of links included in a set at each level, w, and the total number of sets, m_T, are given in Table 1. These may both be computed as a function of the level number and are given as follows:

$$w = 2^{\ell} \tag{19}$$

and

$$m_T = (N+1)/w = 2^{(\ell_T - \ell)} \tag{20}$$

where ℓ is the level number.

From the above definitions, the equations to compute the position/

orientation of the base with respect to the end-effector may be written. Let k and j indicate the numbers of the outside links in a given set, and let i be the number of the middle link of a set. The equations are then given as follows:

For $\ell = 1, 2, \ldots, \ell_T$:

$$\left.\begin{aligned} k &= w\,m && (21)\\ i &= w\,(m - 0.5) && (22)\\ j &= w\,(m - 1) && (23)\\ {}^{k}U_j &= {}^{k}U_i\,{}^{i}U_j && (24)\\ {}^{k}\boldsymbol{t}_j &= {}^{k}\boldsymbol{t}_i + {}^{k}U_i{}^{i}\boldsymbol{t}_j && (25) \end{aligned}\right\} m = 1, 2, \ldots, m_T.$$

Note that the m_T pairs of equations indicated by Eqs. (24) and (25) may be computed in parallel. That is, the indexing on m in this case does not indicate a recursive relationship.

In the Jacobian algorithm, it is necessary to compute the position/orientation of all of the links ${}^{N+1}\boldsymbol{t}_j$ and ${}^{N+1}U_j$, not just the position/orientation of the base with respect to the end-effector. In order to obtain the desired result, the indices on the previous equations may be changed slightly so as to include other links in a set besides the middle one. The resulting equations for the parallel algorithm for obtaining the Jacobian are given as follows:

For $\ell = 1, 2, \ldots, \ell_T$:

$$w = 2^{\ell} \qquad (26)$$

$$\left.\begin{aligned} k &= w\,m && (27)\\ i &= w\,(m - 0.5) && (28)\\ j &= w\,(m - 1) && (29)\\ {}^{k}U_{j+n} &= {}^{k}U_i\,{}^{i}U_{j+n} && (30)\\ {}^{k}\boldsymbol{t}_{j+n} &= {}^{k}\boldsymbol{t}_i + {}^{k}U_i{}^{i}\boldsymbol{t}_{j+n}. && (31) \end{aligned}\right\} \begin{aligned} m &= 1, 2, \ldots, 2^{(\ell_T - \ell)} \\ n &= 0, 1, \ldots, (2^{\ell-1} - 1) \end{aligned}$$

For i = 1,..., N:

$$^{N+1}\boldsymbol{\gamma}_i = {}^{N+1}U_{i-1}\begin{bmatrix}0\\0\\1\end{bmatrix}, \qquad \text{revolute joint;} \tag{32}$$

$$^{N+1}\boldsymbol{\gamma}_i = \boldsymbol{0}, \qquad \text{sliding joint;} \tag{33}$$

$$^{N+1}\boldsymbol{\beta}_i = {}^{N+1}\boldsymbol{\gamma}_i \times (-{}^{N+1}\boldsymbol{t}_{i-1}), \qquad \text{revolute joint;} \tag{34}$$

$$^{N+1}\boldsymbol{\beta}_i = {}^{N+1}U_{i-1}\begin{bmatrix}0\\0\\1\end{bmatrix}, \qquad \text{sliding joint.} \tag{35}$$

Note that the ith column of the Jacobian ($\boldsymbol{\gamma}_i$ and $\boldsymbol{\beta}_i$) may be computed as soon as the position and orientation of the $(i-1)^{st}$ link are

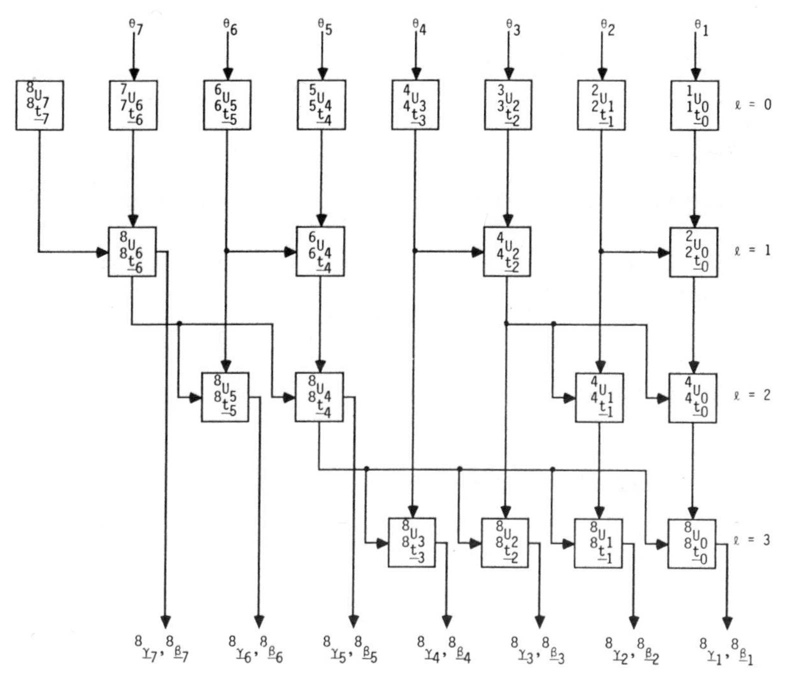

FIGURE 4 Block diagram Showing the Flow of Data in the Parallel Algorithm (N = 7).

computed. That is, there is no preferred sequence on the index i in Eqs. (32) through (35). Also, the indices m and n are completely independent and all combinations of these should be used at each level.

The flow of data through the basic ℓ_T levels of processing may best be shown through a block diagram and this is given in Fig. 4 for a seven degree-of-freedom manipulator. As shown, preceding the required three levels of processing is an initial level ($\ell = 0$) to compute ${}^kU_{k-1}$ and ${}^k\boldsymbol{t}_{k-1}$ for each of the links. After $\log_2(N+1)$ levels, the computation is complete. While the total delay may be significantly decreased through the parallel approach, it should be noted that the communication paths indicated in Fig. 4 indicate non-local communication, and this may cause difficulties when considering systolic array implementation.

SYSTOLIC ARCHITECTURES

The pipeline and parallel algorithms developed in the previous section provide the basis for configuring a systolic array of processors to implement the Jacobian. Figs. 3 and 4 show the flow of data in both the pipeline and parallel algorithms and these suggest specific configurations to be evaluated.

For the pipeline algorithm, if one processor is assigned to each stage of the pipeline, then each processor performs the computation associated with a particular link. If the requirements for computing the orientation, ${}^{N+1}U_{i-1}$, are approximately equal to those of computing the position, $\boldsymbol{r}_{i-1}$, and all other quantities, then the reservation table shown in Fig. 5 (for N = 4) may be used to schedule the pipeline. For the table, $X_i(T_j)$ is the computation involved in computing the orientation, ${}^{N+1}U_{i-1}$, from a sample of the joint angle θ_i taken at discrete time T_j, Y_i applies to the computation of the position, $\boldsymbol{r}_{i-1}$, and Z_i applies to the computation of the Jacobian elements, $\boldsymbol{\gamma}_i$ and $\boldsymbol{\beta}_i$.

The compute time is just the delay from sampling the joint angles to output of the Jacobian and is equal to (N+1)/2 time units. A time unit is the time it takes to perform all computations for a link. The

Processor \ Time	0	0.5	1.0	1.5	2.0	2.5
1				$X_1(T_0)$	$Y_1(T_0), Z_1(T_0)$	
2			$X_2(T_0)$	$Y_2(T_0), Z_2(T_0)$	$X_2(T_1)$	
3		$X_3(T_0)$	$Y_3(T_0), Z_3(T_0)$	$X_3(T_1)$	$Y_3(T_1), Z_3(T_1)$	
4	$X_4(T_0)$	$Y_4(T_0), Z_4(T_0)$	$X_4(T_1)$	$Y_4(T_1), Z_4(T_1)$	$X_4(T_2)$	

Compute Time = 2.5 = (N+1)/2 time units

Initiation Rate = 1.0 (time units)$^{-1}$

CPU Utilization = 100%

FIGURE 5 Reservation Table for One Processor per Link for the Pipeline Algorithm (N = 4 shown). (X_i — orientation computation of link i, Y_i — position computation for link i, Z_i — computation of the Jacobian elements.)

initiation rate is just the sampling rate and may be computed from the following equation:

$$\text{Initiation rate} = \frac{1}{T_{j+1} - T_j} = 1.0\ (\text{time units})^{-1}. \tag{36}$$

The CPU utilization is 100% once the pipe is filled if the computation of the orientation is assumed to be exactly one-half of the total computation for each link. In general, there will be an imbalance between the parts of the total computation required, and this will reduce the CPU utilization to less than 100%.

A number of other pipeline configurations are possible depending on the total number of processors assigned to do the Jacobian computation. Fig. 6 shows the reservation table when two processors are assigned to do the computation for a link. Processor i computes the orientation of link (i−1) while processor (N+i) performs all other computation for link (i−1). The compute time is the same as the N-processor case. However, the initiation rate has doubled from the previous case. The CPU utilization is the same as before and is assumed to be close to 100%.

Fig. 7 gives the reservation table for the case of two processors, one to perform the orientation computation for all links and one to do all other computation. The compute time is the same as the other cases while the initiation rate has been significantly reduced. If a

Processor \ Time	0	0.5	1.0	1.5	2.0	2.5
1				$X_1(T_0)$	$X_1(T_1)$	
2			$X_2(T_0)$	$X_2(T_1)$	$X_2(T_2)$	
3		$X_3(T_0)$	$X_3(T_1)$	$X_3(T_2)$	$X_3(T_3)$	
4	$X_4(T_0)$	$X_4(T_1)$	$X_4(T_2)$	$X_4(T_3)$	$X_4(T_4)$	
5					$Y_1(T_0), Z_1(T_0)$	
6				$Y_2(T_0), Z_2(T_0)$	$Y_2(T_1), Z_2(T_1)$	
7			$Y_3(T_0), Z_3(T_0)$	$Y_3(T_1), Z_3(T_1)$	$Y_3(T_2), Z_3(T_2)$	
8		$Y_4(T_0), Z_4(T_0)$	$Y_4(T_1), Z_4(T_1)$	$Y_4(T_2), Z_4(T_2)$	$Y_4(T_3), Z_4(T_3)$	

Compute Time = 2.5 = (N+1)/2 time units

Initiation Rate = 2.0 (time units)$^{-1}$

CPU Utilization ≈ 100%

FIGURE 6 Reservation Table for Two Processors per Link for the Pipeline Algorithm (N = 4 shown).

Processor \ Time	0	0.5	1.0	1.5	2.0	2.5
1	$X_4(T_0)$	$X_3(T_0)$	$X_2(T_0)$	$X_1(T_0)$	$X_4(T_1)$	
2		$Y_4(T_0), Z_4(T_0)$	$Y_3(T_0), Z_3(T_0)$	$Y_2(T_0), Z_2(T_0)$	$Y_1(T_0), Z_1(T_0)$	

Compute time = 2.5 = (N+1)/2 time units

Initiation Rate = 0.5 = 2/N (time units)$^{-1}$

CPU Utilization ≈ 100%

FIGURE 7 Reservation Table for Two Processors Total for the Pipeline Algorithm (N = 4 shown).

single processor is used, then the compute time is N while the initiation rate is 1/N.

The results for the pipeline case are shown in Table 2. Under the assumptions made, the initiation rate is proportional to the number of processors used. However, the compute time is constant for configurations with more than one processor. This results because two processors allow the orientation and all other computation to be performed in parallel, therefore exploiting all parallelism in the problem. No further gains in compute time will be made by the use of additional processors.

Number of Processors	Initiation Rate	Compute Time
1	1/N	N
2	2/N	(N+1)/2
N	1	(N+1)/2
2N	2	(N+1)/2

Table 2 Initiation Rate and Compute Time Results for the Pipeline Architecture as a Function of the Number of Processors Used. (A Time Unit is the Time Needed to Perform the Computation Associated with One Link.)

With the pipeline algorithm, the only way that the compute time could be further decreased is if the computation were broken down into smaller entities. If the computation were broken down into basic matrix-vector operations, then the parallelism in the problem could be further exploited. However, this would imply considerable communications bandwidth between processors, and this will only be explored in the context of an actual processor design (see the implementation section of the chapter).

The compute time may be further reduced if the parallel algorithm of the previous section is used. A reservation table for the case of N processors is given in Fig. 8. Each processor, as before, is assigned the computation for one of the links. If the computation of the orientation and position across sets of links is assumed to dominate the total computation required, then the compute time is reduced to the order of $\log_2(N+1)$.

As mentioned in the previous section, the communications between processors in the parallel case is non-local and does not lend itself to systolic array implementation. If some redundant computation is permitted in the systolic array, then the parallel algorithm can be mapped onto a hypercube [15]. This is shown in Fig. 9. (N+1) processors are used. A hypercube network is configured between the processors so that processors whose addresses differ by a single bit are considered to be adjacent. The figure indicates that

Processor \ Time	0 – 1.0	1.0 – 2.0	2.0 – 3.0	3.0 –
1	${}^2U_0, {}^2\underline{t}_0$	${}^4U_0, {}^4\underline{t}_0$	${}^8U_0, {}^8\underline{t}_0$	$\underline{\gamma}_1, \underline{\beta}_1$
2		${}^4U_1, {}^4\underline{t}_1$	${}^8U_1, {}^8\underline{t}_1$	$\underline{\gamma}_2, \underline{\beta}_2$
3	${}^4U_2, {}^4\underline{t}_2$		${}^8U_2, {}^8\underline{t}_2$	$\underline{\gamma}_3, \underline{\beta}_3$
4			${}^8U_3, {}^8\underline{t}_3$	$\underline{\gamma}_4, \underline{\beta}_4$
5	${}^6U_4, {}^6\underline{t}_4$	${}^8U_4, {}^8\underline{t}_4$		$\underline{\gamma}_5, \underline{\beta}_5$
6		${}^8U_5, {}^8\underline{t}_5$		$\underline{\gamma}_6, \underline{\beta}_6$
7	${}^8U_6, {}^8\underline{t}_6$			$\underline{\gamma}_7, \underline{\beta}_7$

Compute Time $\simeq 3.0 = \log_2 (N+1)$ time units

Initiation Rate $\simeq 1/3.0$ (time units)$^{-1}$

CPU Utilization $\simeq \frac{21-9}{21} \times 100\% = 57.1\%$

FIGURE 8 Reservation Table for the Parallel Algorithm (N = 7 shown). (The Computation of the Jacobian Elements, γ_i, and $\boldsymbol{\beta}_i$, is Assumed to be Relatively Small.)

after the first time step, communication is between processors whose addresses differ in the least significant bit. After the second time step, communication is between processors whose addresses differ by the second bit, etc.

Because of the communication properties of the hypercube, some redundant computation is required as shown in Fig. 9. However, the compute time is still on the order of $\log_2(N+1)$. However, the CPU utilization has increased to 75%.

In discussing several of the previous configurations, remarks have been made about the difficulties of communications between processors in the systolic arrays considered. A specific processor has been designed at The Ohio State University for robotics and its characteristics will be outlined in the next section. In the section following that, a more detailed evaluation of the compute time, initiation rate, and CPU utilization, for the various configurations proposed above, will be given. It will include a finer breakdown of the computations

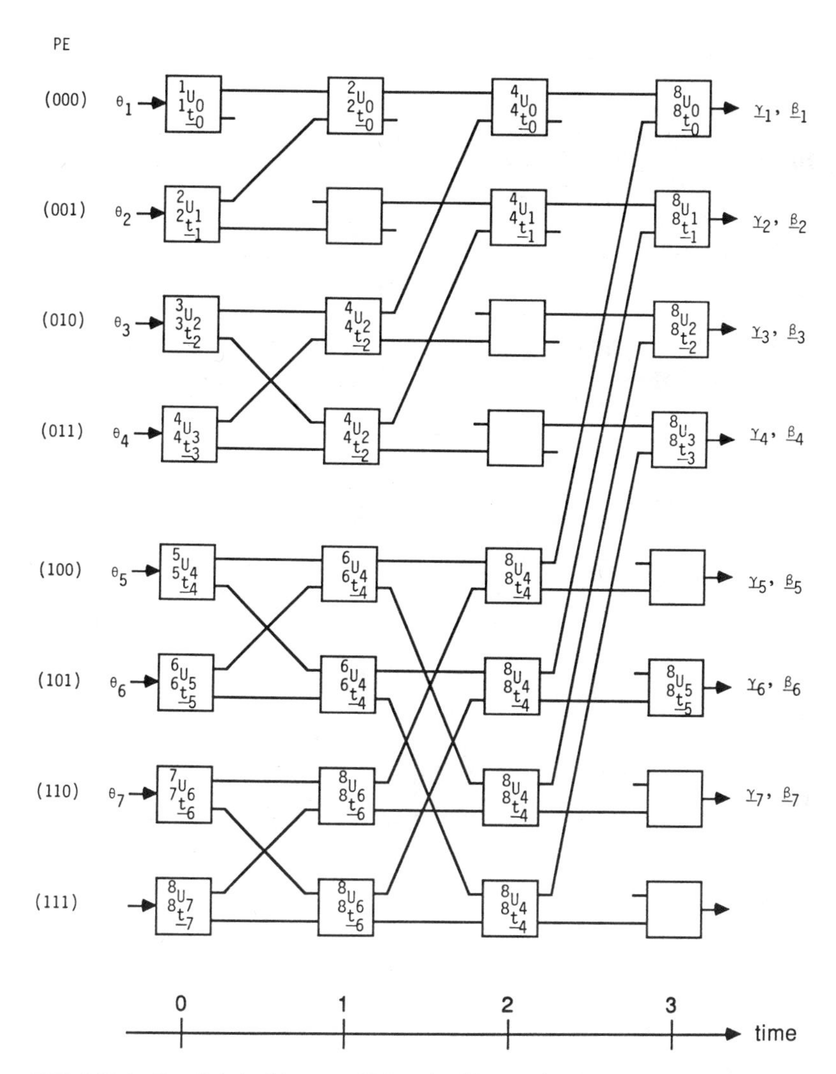

FIGURE 9 Parallel Architecture Using the Hypercube Interconnection Network (N = 7 shown). (PE = Processing Element.)

in order to consider greater amounts of parallelism. Also, the limitations of I/O bandwidth will be fully considered in the analysis.

ROBOTICS PROCESSOR

In order to provide the basic processing element for systolic array implementation of the most common robotics computations, the design for a Robotics Processor VLSI chip has been developed. Fig. 10 gives a block diagram of the Robotics Processor showing the register file (RF), arithmetic units, input/output structure, and data paths. The basic computational units are the floating point adder (FPA) and the floating point multiplier (FPM). These operate in parallel on 32-bit floating point numbers (IEEE format) and have

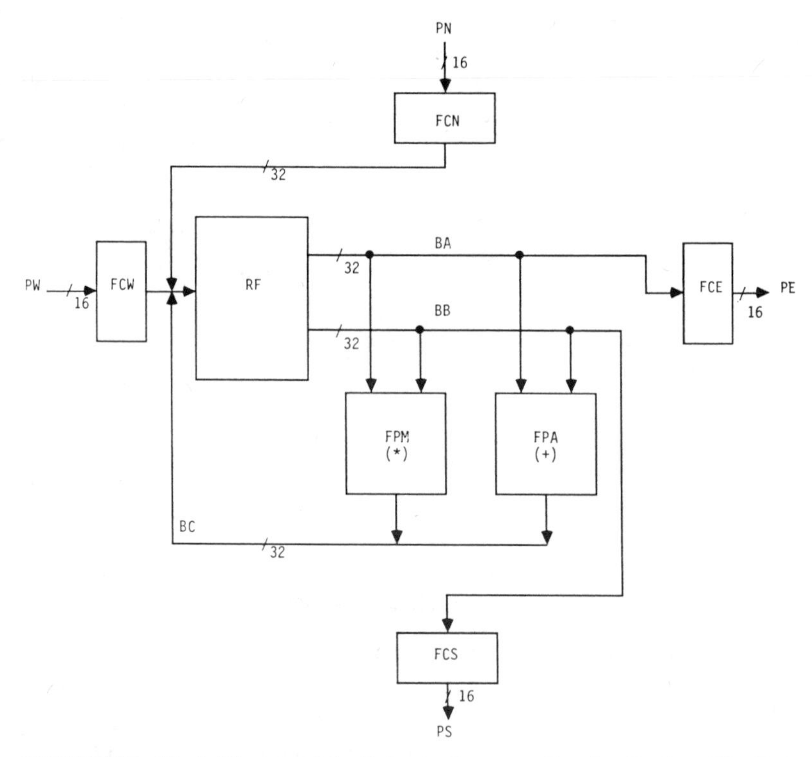

FIGURE 10 Block Diagram of the Robotics Processor Showing Register File (RF), Floating Point Multiplier (FPM) and Adder (FPA), Four Unidirectional Ports (PN, PS, PE, PW), and Three Internal Buses (BA, BB, BC).

three stages of internal pipelining. (Four clock periods are required to fetch operands from the register file and return the result of the operation to the register file.)

The register file consists of up to 128 words of 32 bits each. It is a triple port memory; that is, buses BA, BB, and BC may be used simultaneously. Also, data is multiplexed across each of the buses during a basic clock cycle so that fetching of operands and storing of all results may be accomplished in a clock period (1 μs.).

Four unidirectional ports provide input data and furnish output data to other processors. These are all 16-bit ports and these provide a path for transferring 32-bit data between processors in two clock cycles (from the register file of one processor to the register file of another). A Format Converter (FC) between an internal bus and external port allows transparent access of 32-bit data. The north and west ports (PN and PW) may be used for input while the south and east ports (PS and PE) may be used for output.

The register file and floating point units are designed to operate at a 1 Mhz. rate while the system clock will operate at 16 Mhz. The controller for the processor is microprogrammed. During initialization the microcode, for a particular generic robotics computation such as the Jacobian, is loaded into the control memory (not shown in Fig. 10). When microprogram development is complete, the control RAM could be replaced by an appropriate control ROM.

The Robotics Processor is an example of a Reduced Instruction Set Computer (RISC). It is based on a Harvard architecture [16] in which the program memory and the data memory are disjoint. There are four different types of instructions, each a maximum of 46 bits wide (perhaps smaller if the size of the register file can be reduced thus requiring smaller register addresses). The format for each of the instruction types is given in Fig. 11. The instruction format is an example of a horizontal microcode since very little decoding of the instruction operands is necessary.

The arithmetic instruction allows the floating point multiplier and adder/subtractor units to operate in parallel. Six register file addresses are furnished as a part of the instruction so that all operands and results may be loaded/stored in one clock period. The I/O instruction allows simultaneous input/output from all four unidirectional ports. No means for handshaking is provided so that communication between processors must be carefully synchronized.

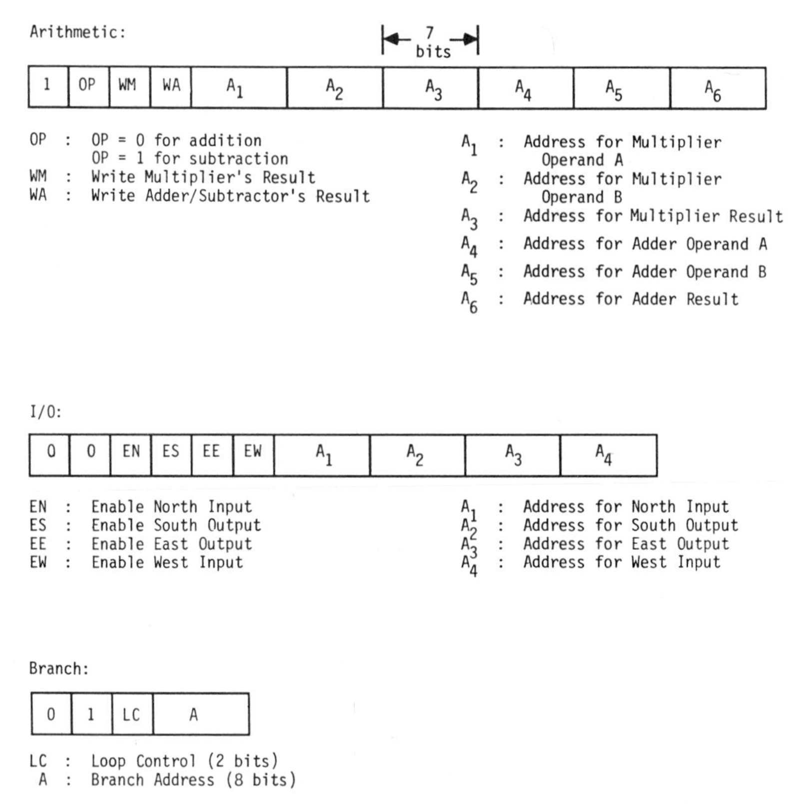

FIGURE 11 Microinstruction Format for the Robotics Processor.

Also provided is a branch instruction. The branch instruction is especially useful in providing looping when the data for several links is processed by one Robotics Processor. A special case of the I/O instruction is all zero's which acts like a NO-OP instruction.

A preliminary design for the Robotics Processor has been made and its feasibility for VLSI implementation has been evaluated [12]. The total number of transistors required is approximately 60K. Using 3 micron NMOS technology, it is possible to fabricate this processor on a moderate die size. (One existing example is the Berkeley RISC II CPU with the same technology, containing about 40K transistors, and having a die size of 171 mil × 304 mil [17].)

IMPLEMENTATION WITH THE ROBOTICS PROCESSOR

Using the Robotics Processor as the basic processing element, a more detailed evaluation of the various systolic array configurations may be made. In particular, the communications overhead will be fully accounted for. Also, there is considerable parallelism in the Robotics Processor itself and this will be used to full advantage. Considerable overlap of the matrix-vector operations required in computing the Jacobian may speed up the calculations significantly so that few processors may be needed.

$$\begin{bmatrix} x_{11} & x_{12} & x_{13} \\ x_{21} & x_{22} & x_{23} \\ x_{31} & x_{32} & x_{33} \end{bmatrix} * \begin{bmatrix} y_1 \\ y_2 \\ y_3 \end{bmatrix} = \begin{bmatrix} z_1 \\ z_2 \\ z_3 \end{bmatrix}$$

$A = x_{11} * y_1$ $\quad B = x_{12} * y_2$ $\quad C = x_{21} * y_1$

$D = x_{22} * y_2$ $\quad E = x_{31} * y_1$ $\quad F = x_{32} * y_2$

$G = x_{13} * y_3$ $\quad H = x_{23} * y_3$ $\quad I = x_{33} * y_3$

$J = A + B$ $\quad K = C + D$ $\quad L = E + F$

$z_1 = G + J$ $\quad z_2 = H + K$ $\quad z_3 = I + L$

Stages \ Clock Cycles	1	2	3	4	5	6	7	8	9	10	11	12	13	14	15	16	17	18
Multiplier Stage 1	A	B	C	D	E	F	G	H	I									
Multiplier Stage 2		A	B	C	D	E	F	G	H	I								
Multiplier Stage 3			A	B	C	D	E	F	G	H	I							
Multiplier-store result				A	B	C	D	E	F	G	H	I						
Adder Stage 1						J		K		L	z_1	z_2		z_3				
Adder Stage 2							J		K		L	z_1	z_2		z_3			
Adder Stage 3								J		K		L	z_1	z_2		z_3		
Adder-store result									J		K		L	z_1	z_2		z_3	

FIGURE 12 Reservation Table for the Multiplication of a 3 × 3 Matrix With a 3 × 1 Vector.

Most of the operations in computing the Jacobian are based on manipulation of 3×1 vectors (V) and 3×3 matrices (M). Reservation tables have been used to schedule these operations on the Robotics Processor and the results are given in Table 3. The number of clock cycles required to complete each operation as well as the relative complexity is given. A vector to vector add (V + V) is considered to have a relative complexity of 1, and all other operations are normalized to this value. Also as an example, a reservation table for a matrix-vector multiply (M·V) is given in Fig. 12. Note that full use of the pipelining and parallelism in the Robotics Processor is made.

Using the relative complexity numbers given in Table 3, a task graph for the Jacobian may be drawn and it is given in Fig. 13. The circles in the figure represent the computation required with the number in the circle giving the relative complexity of the computation. The arrows connecting the circles not only indicate the sequence of the computations, but also show the flow of data from one process to another. The number beside the arrow indicates the relative complexity of the I/O communications required if the processes reside on different processors.

Fig. 13 also indicates a reasonable partition of the computation

Matrix-Vector Operation	Symbol	Clock Cycles	Relative Complexity
Vector Add	V + V	6	1
Vector-Constant Multiply	V · C	6	1
Vector Dot Product	V · V	13	2
Vector Cross Product	V × V	13	2
Matrix-Vector Multiply	M · V	17	3
Matrix-Matrix Multiply	M · M	35	6
Transfer a Vector		6	1

Table 3 Computation Times for Typical Matrix-Vector Operations.

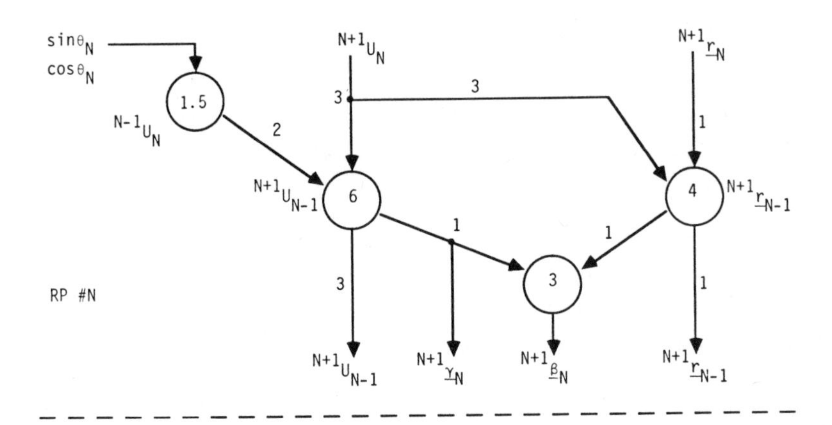

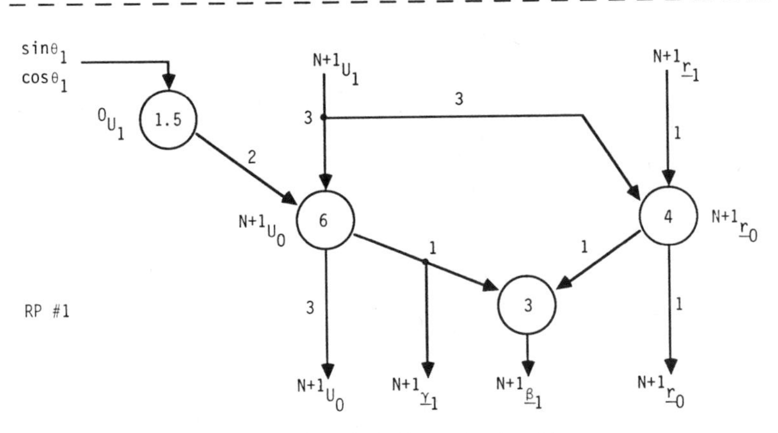

FIGURE 13 Task Graph for Jacobian.

when N processors are used. In this case, the task graph can be very useful as a tool in partitioning the overall computation and scheduling it on processor arrays. With N processors, 14.5 units of computation must be executed on each processor. Also, four units of time are needed to transfer the position and orientation of the

$(i-1)^{st}$ link to the next processor in the pipeline. This indicates something of the overhead involved in communications between processors.

An evaluation of using the Robotics Processor as the basic processing element in a linear systolic array has been made. First of all, the case of N processors configured in a pipeline as shown in Fig. 3 has been considered. Two changes have been made from that shown. Instead of the angle θ_i being input from a host computer external to the array, the sine and cosine of θ_i are furnished. This occurs since the Robotics Processor is unable to compute the sine and cosine functions. Also, input of the position and orientation for a link is from the 'east' and output is to the 'west'. This is directly opposed to that provided by the Robotics Processor. However, there is no real problem since the notions of 'east' and 'west' can be arbitrarily reversed without any actual changes in the design of the array or Robotics Processor.

A microprogram has been written for each of the Robotics Processors to compute all quantities associated with the ith stage of processing. A timing diagram for the schedule of computations and input/output is given in Fig. 14. The individual times shown in the figure are clock periods (1 μs.). Note that a considerable amount of time is devoted to input/output. (Remember that each floating point number takes two clock periods to transfer between processors.) Also, the computation times are less than those expected by using Table 3. This results because full advantage of overlap between matrix-vector operations is made.

It should also be noted that there is some idle time in the schedule of Fig. 14. This occurs because of the need for tight synchronization of input/output between processors.

The total number of registers needed in each processor when N processors are used has been found to be 39. Each of the microinstructions is then 40 bits long. Also the size of the microprogram for each of the processors was 93 words. The total number of bits in the control memory is thus approximately 3.7K. The total memory required including the register file is approximately 4.9K bits.

The total compute time (including input/output and idle time) to completely determine the Jacobian may be derived from the timing diagram and is equal to 48N + 33. (Note that the timing for Processors N and 1 are slightly different than the others since their

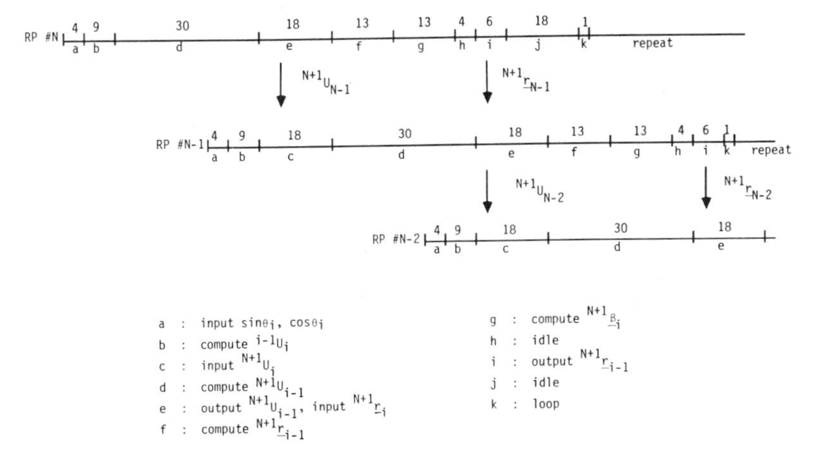

FIGURE 14 Timing Diagram for the Schedule of Computation and Input/Output for N Pipelined Robotics Processors.

input/output requirements are not the same.) The initiation rate may be computed by noting Fig. 14 and is equal to 1/116 (clock periods)$^{-1}$. Also, the processors are idle for four clock periods out of 116 giving a CPU utilization of 97%.

Microprograms have also been written for the case of 1-processor and 2-processor pipeline configurations and (N+1)-processor parallel and hypercube configurations. The results are given in Table IV. As expected, the total compute time decreases less than by a factor of two from the 1-processor to the N-processor pipeline case. Any advantage made in increasing the number of processors to increase the overlap in computation is lost in the increased communications requirements and difficulty of synchronization (increased idle time). Also as expected, the initiation rate goes up substantially with the number of processors used.

The speedup, η, is defined as the ratio of the compute time for a particular case to the compute time for the l-processor case. Note that it is only 1.52 for the N-processor case.

The total number of registers and size of the control memory needed are included in the table since they have a significant impact on the size of the VLSI chip required. The number of registers tends to decrease as processors are added since less storage is needed in

	Processors				
	1	2	N	N parallel	N+1 hypercube
Compute Time (μs)	561	501	369	239	239
	78N+15	68N+25	48N+33	$41+66 \log_2 N+1$	$41+66 \log_2 N+1$
Initiation Rate (μs^{-1})	1/561	1/488	1/116	1/239	1/239
	1/(78N+15)	1/(68N+12)	1/116	$1/(41+66 \log_2 N+1)$	$1/(41 + 66 \log_2 N+1)$
CPU Utilization (%)	100	93	97	64	82
Speedup (η)	1.0	1.12	1.52	2.35	2.35
Number of Registers (32-bit)	87	54	39	39	51
Control Memory Size (bits)	4K	2.7K	3.7K	4K	4K
Total Memory (bits)	6.8K	4.4K	4.9K	5.3K	5.7K

Table 4 Comparison of Pipeline/Parallel Architectures for the Jacobian (N = 7 shown).

each processor. However, the amount of control memory is more difficult to predict since it depends upon a number of factors such as the equations to be implemented by each processor and the input/output overhead.

Also given in Table 4 are the results for the (N+1)-processor parallel configuration and the (N+1)-processor hypercube configuration. The compute times for these are equal and this value is the smallest since it is of the order $\log_2(N+1)$. However, the CPU utilization for each has significantly dropped from 100% since the processors are not fully used in each level of processing.

SUMMARY AND CONCLUSIONS

This chapter has proposed several systolic array configurations for implementing the manipulator Jacobian. These were based on both pipeline and parallel algorithms for the Jacobian which were given earlier in the chapter. The compute time for the pipeline configurations increases with the total number of degrees of freedom of the manipulator, N. Also, increasing the number of processors for implementation of a specific size Jacobian, does not decrease the compute time significantly. The initiation rate, on the other hand, increases at a rate which is close to proportional to the number of processors used.

The compute time for the parallel configurations increases as the $\log_2 N$. However, the CPU utilization is less than that for the pipeline case. Also, the communications required in the parallel configuration is non-local, and thus does not lend itself to systolic array implementation.

The parallel algorithm has also been mapped onto a hypercube configuration and is still characterized by a compute time which varies as $\log_2 N$. It does require some redundant computation so as to eliminate the need for communications between non-adjacent processors, but this may be accomplished during idle periods which are associated with the usual parallel configuration.

If larger numbers of processors are used to implement the systolic array, it is ideally possible to reduce the compute time and increase the initiation rate even further [18]. However, the practical use of increasing numbers of processors is limited since I/O communications difficulties eliminate any gains made in increasing the overlap of computation.

This principle has been established in this chapter in the context of a specific design for a processing element for the systolic array. The design of a VLSI Robotics Processor chip has been used to provide a more detailed evaluation of the various systolic array configurations. The speedup for seven processors (for a seven degree-of-freedom manipulator) in a pipeline configuration was only 1.52. Analysis of this result showed that the problem was directly related to I/O bandwidth limitations. The speedup increased to 2.35 for the parallel configurations, but communications bandwidth was yet a limiting factor.

Results were also given in the chapter for the size of the register file and control memory required in the Robotics Processor chip when implementing the Jacobian. The values obtained indicate that the actual implementation of the chip is within the capabilities of state-of-the-art fabrication facilities.

In the past, use of the Jacobian in advanced control schemes has often been inhibited because of real-time implementation difficulties. Hopefully, the results of this chapter will aid in eliminating this problem so that advanced control schemes for robot manipulators would find wider use.

ACKNOWLEDGEMENTS

The authors wish to thank William W. Schrader for some of the initial ideas in implementing the Jacobian with systolic arrays. Also, our appreciation goes to Dr. Yusheng T. Tsai for assistance with preparation of the manuscript. This research was supported by the National Science Foundation, Computer Engineering Grant No. DMC-8312677.

REFERENCES

1. D.E. Whitney, *Resolved motion rate control of manipulators and human prostheses*, *IEEE Trans. on Man-Machine Systems*, Vol. MMS-**10**, No. 2, pp. 47–53, 1969.
2. C.A. Klein and R.L. Briggs, *Use of active compliance in the control of legged vehicles*, *IEEE Trans. on Systems, Man, and Cybernetics*, Vol. SMC-**10**, No. 7, pp. 393–400, 1980.
3. A. Leigeois, *Automatic supervisory control of the configuration and behavior of multibody mechanisms*, *IEEE Trans. on Systems, Man, and Cybernetics*, Vol. SMC-**7**, No. 12, pp. 868–871, 1977.
4. C.A. Klein and C.H. Huang, *Review of pseudoinverse control for use with kinematically redundant manipulators*, *IEEE Trans. on Systems, Man, and Cybernetics*, Vol. SMC-**13**, No. 3, pp. 245–250, 1983.
5. T. Yoshikawa, *Manipulability of robotic mechanisms*, *Intern. Journal of Robotics Research*, Vol. **4**, No. 2, pp. 3–9, Summer 1985.
6. J.P. Trevelyan, P.D. Kovesi and M.C.H. Ong, *Motion control for a sheep shearing robot*, *Proc. of the First Intern. Symposium on Robotics Research*, Cambridge: M.I.T. Artificial Intelligence Laboratory, 1983.
7. C.H. Wu and R.P. Paul, *Resolved motion force control of robot manipulator*, *IEEE Trans. on Systems, Man, and Cybernetics*, Vol. SMC-**12**, No. 3, May/June 1982.
8. M.H. Raibert and J.J. Craig, *Hybrid position/force control of manipulators*, *Journal of Dynamic Systems, Measurement, and Control*, Vol. **102**, pp. 126–133, June 1981.
9. K.C. Gupta and K. Kazerounian, *Improved numerical solutions of inverse kinematics of robots*, *Proc. of 1985 IEEE International Conference on Robotics and Automation*, St. Louis, Missouri, March 1985, pp. 743–748.
10. S.Y. Oh, D.E. Orin and M. Bach, *An Inverse Kinematic solution for kinematically redundant robot manipulators*, *Journal of Robotic Systems*, Vol. **1**, No. 3, pp. 235–249, 1984.
11. D.E. Orin and W.W. Schrader, *Efficient computation of the Jacobian for robot manipulators*, *Intern. Journal of Robotics Research*, Vol. **3**, No. 4, pp. 66– 75, Winter 1984.
12. H.H. Chao, *Parallel/pipeline VLSI computing structures for robotics applications*, Ph.D. dissertation, The Ohio State University, Columbus, Ohio, June 1985.
13. J. Denavit and R.B. Hartenberg, *A kinematic notation for lower-pair mechanisms based on matrices*, *ASME Journal of Applied Mechanics*, Vol. **23**, pp. 215–221, 1955.
14. R.P. Paul, *Robot manipulators: Mathematics, programming, and control*, Cambridge: The MIT Press, 1981.
15. K. Hwang and F.A. Briggs, *Computer architecture and parallel processing*, New York: McGraw Hill, 1984.
16. H.G. Cragon, *The elements of single-chip microcomputer architecture*, *Computer*, Vol. **13**, No. 10, pp. 27–41, 1980.
17. R.W. Sherburne, M.H. Katevenis, D.A. Patterson, and C.H. Sequin, *32-bit NMOS Microprocessor with a large register file*, *IEEE Journal of Solid-State Circuits*, Vol. SC-**19**, No. 5, October, 1984.
18. R.H. Lathrop, *Parallelism in manipulator dynamics*, *Intern. Journal of Robotics Research*, Vol. **4**, No. 2, pp. 80–102, Summer 1985.

Chapter 4

PIPE[1]: A SPECIALIZED COMPUTER ARCHITECTURE FOR ROBOT VISION

Dr. Ernest W. Kent

Sensory-Interactive Robotics Group
The National Bureau of Standards[2] [3]
Washington DC 20234

1. PIPE is a registered trademark of Digital/Analog Design Associates, Inc. of New York.

2. This work is the product of United States Government employees, and is not subject to U.S. copyright.

3. The author's current address is: Dept. of Robotics and Flexible Automation, Philips Laboratories, 345 Scarborough Rd. Briarcliff Manor, NY 10510

PIPE (Pipelined Image Processing Engine), is an experimental, multi-stage, multi-pipelined image processing device. It can acquire images from a variety of sources, such as analog or digital television cameras, ranging devices, and conformal mapping arrays. It can process sequences of images in real time, through a series of local neighborhood and point operations, under the control of a host device. Its output can be configured for monitors, robot vision systems, iconic to symbolic mapping devices, and image processing computers. In addition to a forward flow of images through successive stages of operations as in a traditional pipeline, other paths between the stages of the device permit concurrent, interacting pipelining of image flow in other directions. In particular, recursive paths returning images into each stage, and feedback of the results of operations from each stage to the preceding stage are supported. The architecture facilitates a variety of functions such as relaxation and interactions of images over time. Numerous operations are supported; within each stage these include arithmetic and Boolean neighborhood and point operations on images. Between-stage operations on each pixel include thresholding, Boolean and arithmetic operations, functional mappings, and a variety of functions for combining pixel data converging via the multiple pipelined image paths. The device also implements alternative processing modes, including "MIMD" operations specific to regions of interest defined by the host device or by previous operations on the image, and variable resolution pyramid operations.

REQUIREMENTS OF ROBOT VISION

A robot vision system has requirements which differ in some respects from those of other types of machine vision. Since the function of a robot is to perform physical actions in space and time similar to those performed by humans, it is not surprising that these requirements are similar to those imposed on biological vision systems. In some machine vision applications, such as interpretation of images from diagnostic or satellite equipment, each new picture is a new problem; in robot vision, like biological vision, the input is a stream of images in which each frame differs from the last by some small displacement of the camera or of the objects in the scene. In many machine vision tasks the important function is to recognize certain kinds of objects, but in robot vision, again like biological vision, most of the vision system's time is spent providing information about the position of things in space for guidance and servoing. This is true after objects have already been recognized, before they are recognized, and even if they cannot be recognized by the system. The robot must understand the spatial occupancy of its environment and its own relation to it in order to avoid striking surfaces, whether or not those surfaces are part of recognized objects. Ideally the robot vision system should provide a description of the geometric properties of unrecognized objects sufficient to permit the robot to manipulate them; for example, to remove a foreign object from the workspace. Most importantly, a robot vision system shares with the biological system the necessity of synchronization with real-time events.

From these considerations, we can derive some important properties of a robot vision system. Because the robot's function is oriented towards actions and events which extend over time, the appropriate unit of analysis is the image sequence rather than a single image. This implies that the system must capture and operate on image sequences of a length appropriate to the speed of events. Because the robot's world possesses continuity, the system should take advantage of information discovered in previous views. This suggests that the system should not only build internal models of the environment from successive views, but also that it should be able to make use of hypotheses from this model in interpreting subsequent images. The importance of servoing over classification implies that the system must provide information at many levels of analysis. That is, it may be required to supply information on range to points, on inclinations of edges and surfaces, on translation and rotation velocities of features, and similar descriptive properties of objects in space and time. These must be made available to the robot control system as rapidly as they are discovered, prior to and independently of classification. Above all, the system must operate in real-time; a late answer is no answer for a robot guidance system. It is this which compels us to examine special-purpose architectures as a means for accomplishing the other criteria.

THE NBS ROBOT VISION SYSTEM

The general plan of a robot system which incorporates all of these functions is easy enough to sketch, and the basic ideas are included in Figure 1, which is adapted from a robot system being developed by the Sensory-Interactive Robotics Group at NBS. On the left is a sensory processing hierarchy, in the center a knowledge representa-

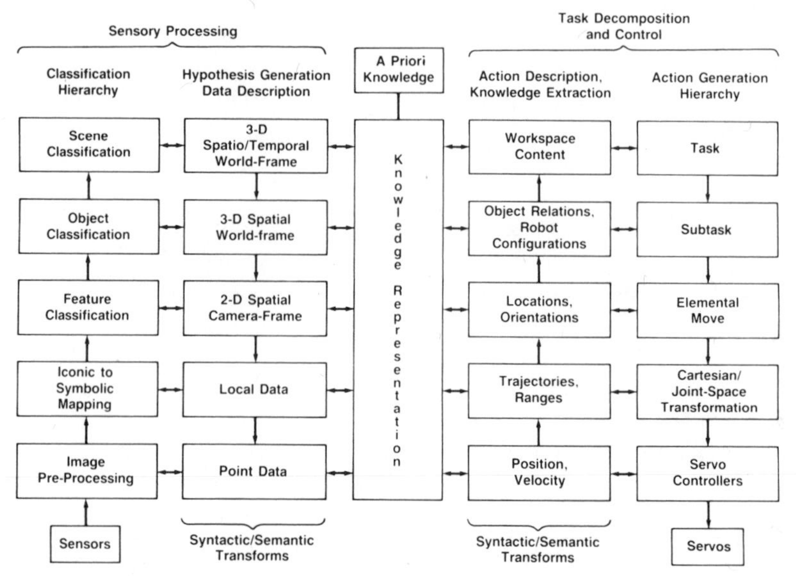

FIGURE 1 The elements of the NBS image-processing configuration in relation to the NBS robot system.

tion system, and on the right a task-decomposition and control hierarchy. This entire system could be considered a 'special purpose architecture', in the sense that it is a special purpose machine which has computers as components. However, most of the elements are currently implemented on individual computers of traditional design, and in this chapter I intend to focus primarily on some particular vision elements within this system which, for reasons of required processing speed, have been implemented on special purpose architectures in the sense of specially-designed hardware.

Before considering these elements in detail, a brief review of the system operations will help clarify their tasks and design goals. The sensory processing hierarchy accepts data from the sensors (in particular cameras for vision sensing) and attempts to form a hierarchical syntactic description of the current view of the world. Note that symbolic and parametric information is made collaterally available to the system's knowledge representation at every level of this process. The task-decomposition and control hierarchy on the opposite side accepts goals in the form of external commands. It attempts to use generic knowledge of how to choose actions together with particular knowledge of the current state of the world to generate actions which accomplish the goals. Knowledge of the current state of the world is made collaterally available to every level of the task-decomposition and control hierarchy from the system knowledge representation.

The knowledge representation contains much more than the information currently available from the sensory processing hierarchy. In addition to information which comes from *a priori* knowledge sources, and knowledge of the control hierarchy's current state, the knowledge representation contains a description of the world built up over all past views from the camera. It thus contains much information which is not extractable from the current viewing window, but which bears on the interpretation of that which is. The current scene descriptions from the sensory processing hierarchy must be reconciled with this information, and the sensory processing hierarchy in turn may be guided by it. Similarly, the knowledge representation's model of the world must be formulated into specific answers to any particular instantaneous requirement of the task-decomposition and control hierarchy. In turn, the state of the action-generating processes provides guidance for the knowledge representation in interpreting its input from the sensory system. There is thus a two-way flow between the knowledge representation and the sensory processing system which "servos" the internal model of the world to the observed data, and which provides hypotheses that guide the processing and interpretation of the data. A similar reciprocal flow links the knowledge representation's internal model of the world with the questions and information produced by the control hierarchy.

The dimension along which the control hierarchy is divided into functional levels (task, subtask, elemental cartesian move, joint-space velocity) is not the same as the dimension along which the vision processing hierarchy is divided into levels (image-plant point properties, image-plane features, world-frame objects.) The knowledge representation, which is not itself hierarchically structured, contains information relevant to all of the different descriptions of the world required by these levels. Its information is maintained in a semantic form from which syntactic and parametric descriptions of the world adapted to the requirements of any of the functional modules can be produced. The representation scheme of this semantic form is chosen principally for convenience in maintaining and organizing the information. The syntactic forms of the various levels of the sensory and control systems are chosen primarily for functional utility in the modules concerned. The "syntactic/semantic transform" modules extract level-specific information from the knowledge representation and instantiate it into the frame and symbols needed for any given functional module. In the other direction, they contain systems for incorporating level-specific syntactic and parametric sensory or control information into the internal model.

This diagram outlines a general conception of the relations of sensory and other knowledge in a robot system. What is vastly more difficult is to specify the algorithms and hardware required to give substance to the boxes in such a diagram. A first generation system was constructed entirely from micro-computers. It used binary image processing of both normal and structured-light images. The knowledge representation and the range of requests acceptable from the control hierarchy were very limited in scope. Nonetheless, this system could accept CAD descriptions of simple parts, recognize them, and provide servo information that allowed the control system to manipulate them in real-time. A second generation system has been under development for some time. This project includes much more elaborate structures for knowledge representation, and a correspondingly richer sensory processing capability which employs gray-scale vision. Most of this second generation system development effort consists of software improvements still running in a multi-microcomputer dataflow (as opposed to distributed) system. With the move to gray-scale vision, however, it became apparent that the early stages of image processing involved computations which could not be handled in real-time by microcomputers. Further, it was felt that for these early image-processing functions a sufficiently good understanding of the requirements of robot vision existed to justify the development of special-purpose vision architectures. Accordingly, an image preprocessor (called PIPE, for Pipelined Image Processing Engine) and a feature extractor (called ISMAP for Iconic to Symbolic MAPper) were designed. The first prototypes of these devices are now in operation. Their relations are depicted in Figure 2, which shows the data-flow paths for the PIPE

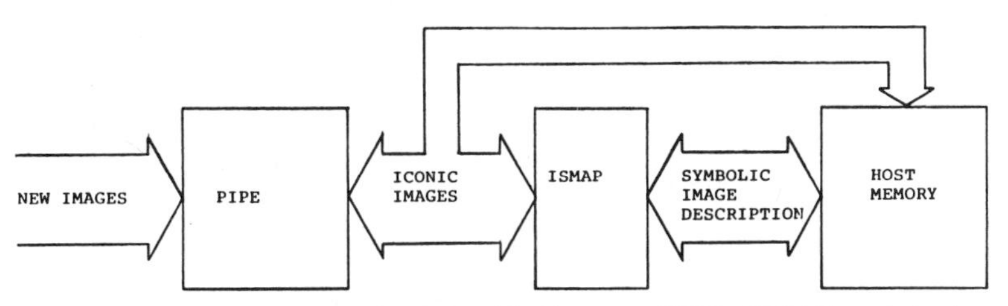

FIGURE 2 Major Image-Flow relationships between PIPE, ISMAP, and the host devices.

and ISMAP elements, to and from the upper levels of the system (indicated here as "host memory").

These devices will replace microcomputers as the first two elements of the sensory processing hierarchy in the second generation of the robot system depicted in Figure 1. The first of these elements is an initial processing stage (image pre-processing) which accepts an image consisting of an array of gray-scale values and produces a similar array of symbolic values which encode features of the gray-scale array, such as edges, or textures. This process of feature detection produces an iconic description of image features from an iconic description of image intensities. That is, for both the input and the output of this process, the global geometric relations of the input values or the output features are implicit in their location in the image array. The feature detection process transforms one iconic array into the other by making local properties explicit as symbols. In most cases the process attempts to describe local properties (features) which achieve some independence of circumstances such as illumination.

Following the feature detection process, the second element of the system performs a feature extraction step (iconic to symbolic mapping). This begins the process of producing a relational feature description of objects which is independent of circumstances of viewing position. In this step, the iconic image of features, in which symbol values are indexed by location, is mapped into a space in which location is indexed by feature value. This produces a representation in which the data which was previously explicit (the feature type) is now implicit in the location of the data in the representation. The image location of the features, which was previously implicit in the iconic representation, is now the explicit data represented by values in the new space. This accomplishes two goals; it permits finding the locations of features of interest by direct indexing rather than by exhaustively searching the image, and it extracts locations as data to which subsequent operations can be applied. These subsequent operations attempt to find meaningful relations among the location data which correspond to geometric relations of features. The geometric relations discovered then serve as input to subsequent classification processes.

DESIGN PHILOSOPHY OF PIPE

In designing the PIPE and ISMAP devices to play these roles in the system, we attempted to optimize the design for the special requirements of robot vision discussed above. The principal goals selected were: 1) real-time processing of images at field-rate, 2) provision for interactions between related images, such as those arising from dynamic image sequences or from stereoscopic views, 3) provision of the ability to apply different algorithms to different regions of the image in real time, 4) ability to perform multi-resolution image processing, and 5) provision for guiding processing by knowledge-based commands and "hypothesis images" supplied from the upper levels of the system.

PIPE is a hardware device specialized for parallel image processing rather than a fully general purpose parallel computer. Through its design, it facilitates a variety of common and important image-processing techniques, as well as several experimental approaches. Within the broad limits of the processes it supports, it is an extremely fast and flexible device. On the other hand, it is not a general purpose computer and it is not possible to program arbitrary algorithms on it, or at least not in an efficient manner. In most cases we have found that processes which are well suited to PIPE may be substituted for others which are not, while accomplishing the same image-processing goal. PIPE is intended as a processor for local operations on images; it is not designed to perform efficiently operations that require global knowledge of the image. It is intended that PIPE operate in conjunction with a host, such as the upper levels of the NBS robot vision system, which will perform global image operations, relieved of the processing burden of large scale repetitive local operations. Thus, PIPE itself accepts iconic data images, and typically produces iconic images whose pixel values are Boolean vectors describing local properties of the pixel neighborhood. PIPE relieves the upper levels of the system of costly low-level local processing which must be performed over the entire image space.

The basic design of PIPE was partly inspired by an earlier proposed machine[1,2] which was in turn adapted from biological models of image processing. The essential organization is a three-dimensional architecture consisting of a sequence of image-processing planes (Figure 3.) Each processing plane has storage arrays which receive images, and operators which act on them. The storage arrays may be considered to be in-register with one another in the third dimension, so that each pixel neighborhood in an array occupies a step in a pipeline of processes extending up through the stack of planes. The result is an image-flow architecture in which the images move upwards through the stack as subsequent images replace them from below. At each stage the operations on the pixels provide interaction in the lateral directions with neighbors in the same array, and in the vertical directions with neighbors in the preceding and succeeding arrays. Logically the machine can be considered as a bundle of bidirectional pipeline processors with lateral interactions.

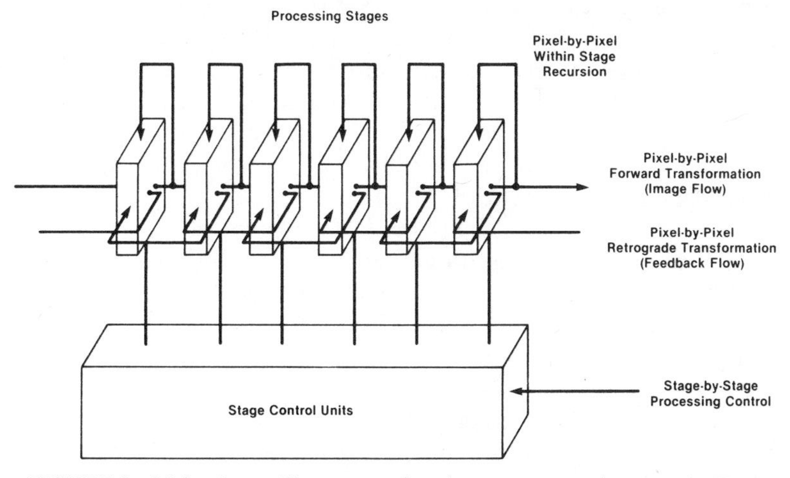

FIGURE 3 Major Image-Flow connections between processing stags in PIPE.

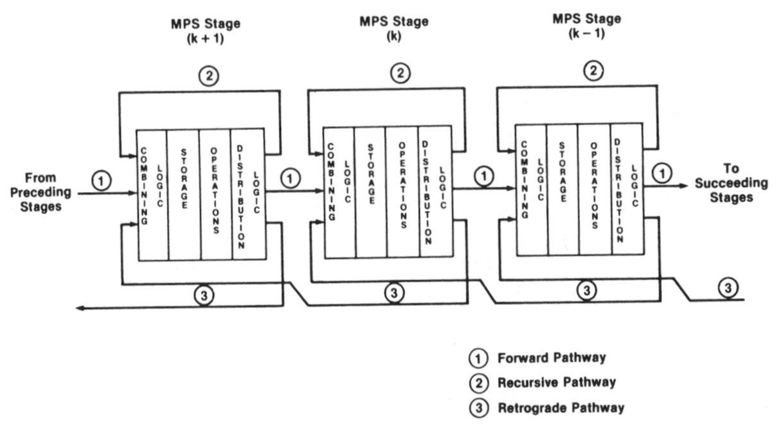

FIGURE 4 Image-Flow connectivity in relation to the major elements of a hardware PIPE stage.

In practice, PIPE actually consists of stages with storage arrays and computational modules (Figure 4) which operate over every pixel neighborhood in the arrays in a single field time (1/60 sec.) Images are transferred from stage to stage at field-rate (60 images/sec) by three concurrent pathways which provide for interacting, image-flow transfers between stages. These interconnect a variable number of identical modular image-processing stages. The three pathways, shown in Figure 3, are: the forward pathway, which acts as a traditional pipelined image-processing path; the retrograde pathway, which carries images in the opposite direction (i.e., from the output of a stage to the input of its predecessor); and the recursive pathway, which carries an image from the output of a stage back into the input of the same stage.

At the input to each stage (labeled "combining logic" in Figure 4) the images carried by the three pathways may be subjected individually to any arithmetic or Boolean operation. Any linear arithmetic or Boolean operation may then be used to combine them into a final input image, prior to its storage in one of two buffers within the stage.

Within each stage, the computational operators may act on images stored in either or both of the two buffers. These operations are contained in the sections marked "operations" in Figure 4. Each stage can perform two simultaneous and independent arithmetic or

Boolean neighborhood operations. Following the neighborhood operations, and prior to output from the stage, the images resulting from the neighborhood operations, or either of the images in the buffers, may undergo a further transformation by an arbitrary function of one or two arguments. If the function is of two arguments, the second argument may be drawn from any source within the stage, including the transformations of same or the other image. The forward, recursive, and retrograde pathways, in any combination, may accept images from the result of any of these operations or from any of the buffers. This is controlled by the section marked 'distribution logic' in Figure 4.

The representation of Figure 4 shows the actual physical grouping of processing elements and storage as they exist in the machine for structural reasons. Functionally, the combining logic of one stage and the operations and distribution sections of the preceding stage may form a more convenient conceptual processing unit for many purposes. Figure 5 shows the architecture of PIPE redrawn in this fashion, where the processors are the conceptual rather than the physical units. In this figure the frame-buffer storage elements can be unbundled from the processing units, so that the tri-directional image-flow relations are easier to visualize.

In an alternative mode, one of the two image buffers in each stage may serve as a map for selecting the processing algorithms being applied to the contents of the other buffer. In this mode, PIPE functions as multi-instruction stream multi-data stream (MIMD) machine, with one buffer defining regions of interest over which each set of algorithms shall be applied, on a pixel-by-pixel basis, as the image is processed. In the prototype version, sixteen such alternative processing algorithms may be selected for different regions of the image within a single field time. However, this is easily increased (up to 256 processing algorithms) simply by adding additional storage to each stage to hold the appropriate tables and parameters.

A third mode permits PIPE to function as a multi-resolution pyramid machine. In this mode, the images carried by the forward pathway are reduced in size by one half at each stage, while sizes of the images carried by the retrograde pathway are doubled at each stage. The images carried by the recursive pathway remain unchanged in resolution. Any combination of stages may operate in this mode, under program control.

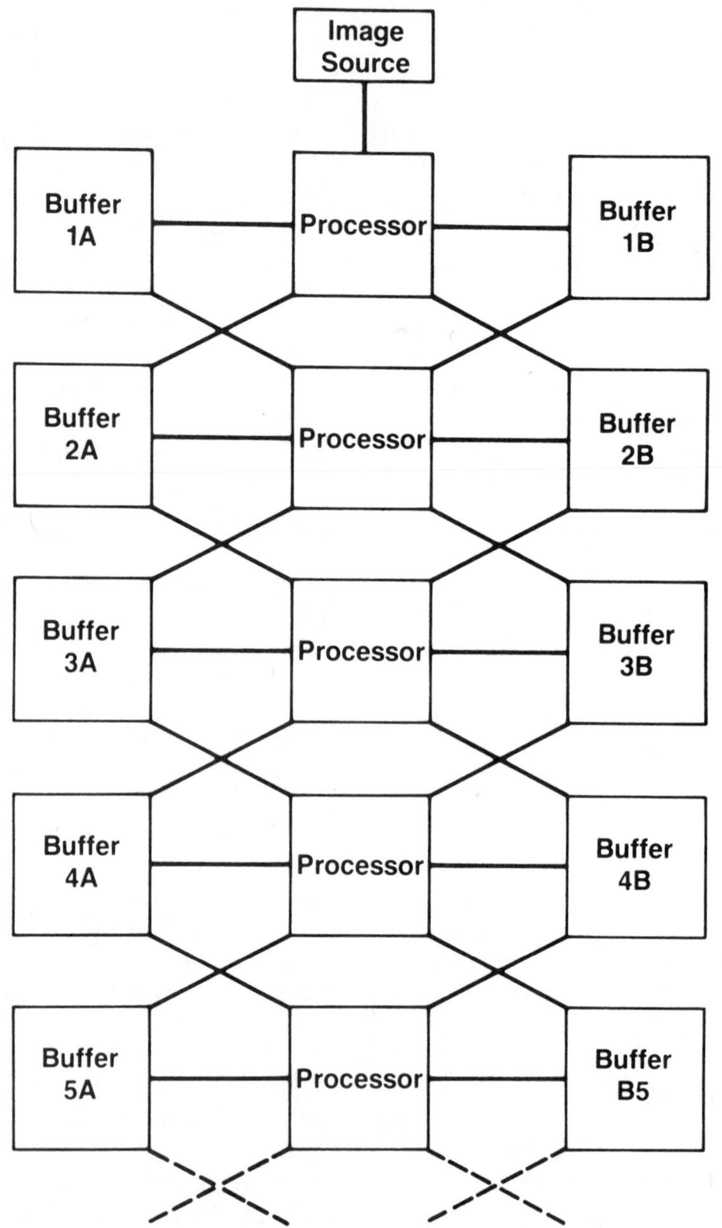

FIGURE 5 Image-Flow connectivity in relation to conceptual functional processors of PIPE.

In addition to the pathways mentioned, PIPE contains four "wild-card" busses which may be used to transport images from higher levels of processing, or from any buffer in the machine, into any other buffer in the machine. It also has two special stages, for input and output, which communicate with the sensors, with processors in upper levels of the system, or with special auxiliary devices. DMA channels allow video display or streaming input and output from any buffer in the machine.

In any of PIPE's operating modes, the operations of every stage are completely independent, and can be completely reconfigured in the inter-field time by an associated stage-sequencer control unit, which in turn may select stage configurations from a stored sequence, or on command from the higher levels.

OPERATION OF PIPE

The initial version of PIPE consists of a sequence of identical image processing stages, sandwiched between a special input processor and a special output processor. The input processor accepts an image from any device that encodes two-dimensional images. It serves as a buffer between the rest of the image processing stages and the outside world. Each successive processing stage receives image data in an identical format, operates on it, and passes it on to the next stage for further processing. This sequence is repeated every television field-time. When an image emerges at the far end of the sequence, it is processed by the special output stage and presented to upper levels of the robot vision system or to a host computer.

The image processing stages between the input and output stages are all identical and interchangeable, but can each perform different operations on the image sequences that they encounter. Usually, each image processing stage receives three input images and transmits three output images. The input images arrive from the processing stage immediately behind each stage, from the processing stage immediately ahead, and from a result of the preceding operation performed by the image processing stage itself. Similarly, the results of processing a current image are transmitted by each stage to the next processing stage in the sequence, to the immediately-preceding processing stage, and recursively back into the image processing stage itself. These three outputs are usually not identical, and each may furnish part or all of the inputs to other stages for the subsequent step in processing. The three inputs may be weighted and combined in each image processing stage, in any fashion, before they are processed.

In addition to these input and output paths, the four "wildcard" paths may be sources or destinations for input and output. Unlike the three principal paths connecting the stages, these wildcard paths are common to all stages, so that only one stage can write to a particular wildcard path at a time, but any or all stages can accept input from them. The wildcard paths allow images to be moved arbitrarily between stages, instead of having to step through from stage to stage. There are no restrictions on the number of destinations for an image output to a wildcard path.

Although there are physically only three pathways between stages (excluding the "wildcard" busses), every pixel neighborhood in an image is processed and sent over these paths in every field time. The result is that PIPE simulates a fully parallel image-flow machine; each pixel appears to have a real-time, private line to an homologous pixel processor in three target stages. There are numerous reasons for requiring the three input and output paths from each image processing stage. It is clear that the forward path allows a chain of operations to be performed on sequential images, giving rise in real time to a transformed image stream (with a constant delay). Similarly, the recursive path allows a pipeline of arbitrary length to be simulated by any stage. It also facilitates the use of algorithms that perform many iterations before converging to a desired result (e.g., relaxation algorithms, or the simulation of large neighborhood operators by successive applications of smaller neighborhood operators). The path to the preceding image processing stage allows operations to be performed using temporal as well as spatial neighborhoods. It also allows information inserted at the output stage by the upper levels of the system to participate in the processing directly. This, for example, allows expectations or image models to be used to guide the processing at all levels, on a pixel-by-pixel basis.

It is helpful in understanding the functions of these processing pathways to consider each in isolation first. If only the forward input path is operative (i.e., the combining weights for the retrograde and recursion paths are set to zero), we have a simple image pipeline processor which can sequentially apply a variety of neighborhood operators to the series of images flowing through it. It can perform either arithmetic or Boolean neighborhood operations and, by thresholding, convert an arithmetic image into a Boolean image. For example, it might be used to smooth an arithmetic gray scale image, apply edge detection operators to it, threshold the "edginess" value to form a binary edge image and then apply Boolean neighborhood operations to find features in the edges. The operation types and parametric values for these operations would be set individually for each stage by the stage control units, which in turn would be instructed (for example from the upper levels of the system) via the input marked "stage-by-stage processing control" in Figure 3.

A second single-path case results if both the forward and retrograde paths' combining functions are zero. Assume that images had previously been loaded into all the processing stages. The recursive path would then cause the image field in each stage to pass through the forward or backward transformation operation recursively, while the images "marched in place". A variety of relaxation operations can be implemented in this way.

For the final single path case, consider that the weights assigned to the forward and recursive paths are zero, leaving only the retrograde pathway active. When the set of such paths is considered in isolation, it becomes clear that it forms a processing chain that is a retrograde counterpart of the forward pipeline. It would, in fact be possible to select appropriate retrograde transformations, insert fields of data at the back of the device, process them through to the front, and get the same result as running the system in the normal direction. The purpose of this is not to provide a bidirectional image processor, but to permit input (at the "output" end of the device) of synthesized images. Such images influence the processing of the normally flowing images by direct interaction, and correspond to "expectancies", "models", "hypotheses", or "attention functions."

The retrograde images are not only able to affect processing of the forward images, but are affected themselves by interaction with them. (The effects that the two image sequences exert on each other may be different because the neighborhood operators on the for-

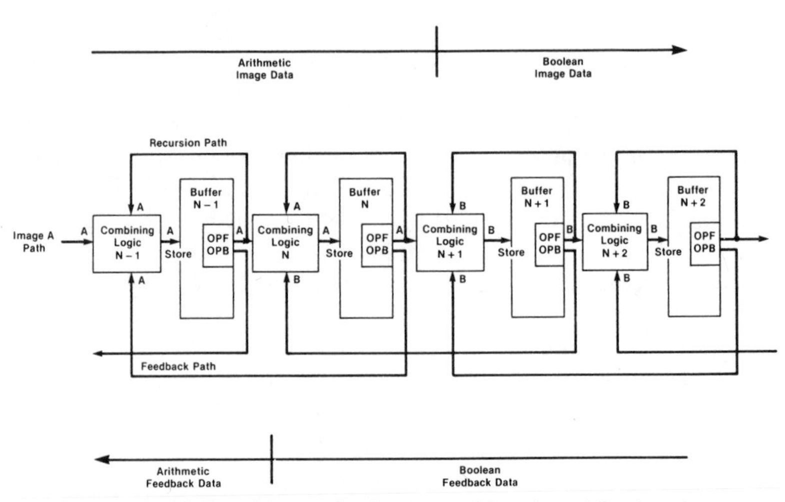

FIGURE 6 Modes of interaction between arithmetic and Boolean images.

ward and backward paths are independent). Retrograde images will usually be generated by knowledge-based processes in higher level components of the robot vision system. They may initially appear in Boolean form, but, as shown in Figure 6, provision is made for all four possible combinations of arithmetic and Boolean inputs and outputs in the combining logic between stages. This permits a descending Boolean image to be instantiated into arithmetic image values by interaction with the ascending arithmetic image. This occurs in the same stage in which the ascending arithmetic image representation is thresholded to become a Boolean image (Stage "N" of Figure 6.) Both the ascending data image and the descending "hypothesis" image can pass across this interface. A major function of PIPE will be to explore the effectiveness of various approaches to hypothesis-guided iconic image processing.

There are a great many ways in which various combinations of these pathways can be used in processing by operating on combinations of processed images arriving over the three pathways. I will consider only a few illustrative examples.

If the preceding and succeeding image fields are considered to contain future and past instances of a field, respectively (as is true in a dynamic image), then forward path corresponds to a path from the future, recursion to a path from the present, and the retrograde path to one from the past. The weighted sum of the three paths' contents the forms a convolution operation on the temporal neighborhood of a pixel. This may occur at the same time as a spatial neighborhood convolution operation is being performed on the contemporary spatial neighborhood in each stage, so that a combined spatio-temporal convolution is achieved.

Boolean information can be processed in an interesting way by combining the outputs of the forward operator from the previous stage and the recursive input from the current stage. Consider the case of a single stage treated in this fashion for eight field-times, using "SHIFT recursive then OR recursive with forward" as the combining operation. Let the incoming images from the previous stage (forward path) have Boolean values resulting from thresholding of eight successive and independent feature detection operations such as oriented edge detection, texture measures, etc. The second stage will accumulate images from the eight preceding Boolean operations into an image composed of eight-bit Boolean vectors, each bit representing the presence or absence of an independent image property at that location. Subsequent Boolean neighborhood operations may then apply independent operators to each bit plane of a neighborhood of such vectors.

HARDWARE DETAILS OF THE PIPE STAGES

Input Stage

A special input stage is used to capture and buffer images from input devices. This allows PIPE to accept digital or analog signals from any device using standard RS-170 television signals and timing. Either of two selectable analog signals is digitized by an eight-bit real-time digitizer. The input stage is capable of acquiring a digitized image of 256 × 240 pixels while remaining synchronized with RS-170 signals. Alternatively, it can capture 256 × 256 pixel images from non-RS-170 signals while internally employing non-standard pixel rates. It can continually capture such images at standard television field rates, and place them in either of the two field buffers contained in the input stage. While storing an image into one of these buffers the input stage can also simultaneously store an image into either buffer, such as a difference image, formed by an ALU operation between the incoming image and a previously captured image. The contents of either of the buffers in the input stage can be sent to the first of the processing stages, while the next image is being acquired.

PIPE accepts eight-bit input data, and this precision is maintained throughout the machine. Intermediate arithmetic operations within subsequent stages are carried to sufficient precision to insure no loss of accuracy when the result is rounded to eight bits for transmission to subsequent stages. The data may be treated as either unsigned eight-bit numbers, or as two's complement signed numbers with the high bit indicating the sign, or as boolean vectors of eight independent bits. All stages may independently select the representation employed, so that unsigned input data may be processed until an operation which generates negative values occurs, treated as signed data thereafter until a thresholding operation occurs, and then treated as Boolean data.

Processing Stages

Following the Input stage, there are a series of modular processing stages (MPS). The MPSs are the "stages" referred to in the preceding sections, and are the elements which perform most of PIPE's processing. All MPSs are of identical modular construction, and are physically interchangeable simply by switching circuit boards. Thus, any MPS can operate at any position in the processing chain, and the processing chain can have a variable length. Eight MPSs are employed for the present development phase of PIPE. The block diagram of a MPS is portrayed in figure 7.

The pre-storage input section of the Nth MPS accepts three eight-bit 256×256 pixel images as input. These come from the forward output of the N-1st MPS, from the recursive output of the operation performed on the previous contents of the Nth MPS, and from the retrograde output of the N+1st MPS. Each data stream may consist, independently of the other two, of arithmetic or Boolean (eight-bit Boolean vector) data, but a given data stream entering a MPS must be entirely Boolean or arithmetic within any single image field.

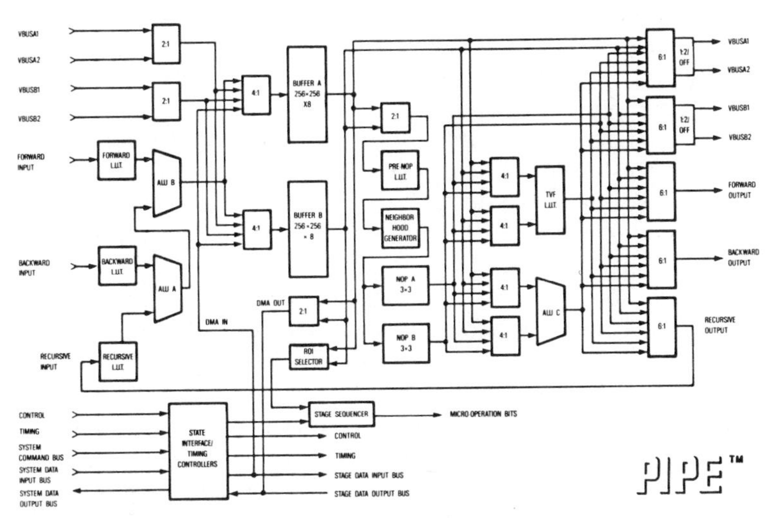

FIGURE 7 The internal architecture of a PIPE processing stage.

Before generating a final eight-bit image from the three data streams, the input section of each MPS performs an arbitrary table-lookup transformation on each them independently and simultaneously (forward, backward, and recursive L.U.T. in Figure 7.) Each of the three lookup tables is one of three sets of 32 tables selectable by the stage operation micro-instruction. The resulting three Boolean and/or arithmetic data streams are then combined through independently-programmable full-function ALUs into a single arithmetic or Boolean data stream (Figure 7, ALU-A and ALU-B.) This data stream is then used to load either of the two selectable field buffers within the MPS (buffer A and buffer B in the figure.) Alternatively, either or both buffers can be filled using the wildcard busses or by direct DMA from the host. The contents of both of these field buffers are then available to subsequent operations of the MPS. External device access to the data in these buffers is also available; an external device may read from or write into either buffer in a random access manner at 400,000 pixels/sec., with auto-indexed addressing supported on command. The wildcard busses provide streaming access to external devices (including monitors) at pixel rates.

The hardware that implements the output functions of the MPS, subsequent to the field buffer storage step, is physically contained on a separate circuit card to allow it to be replaced with other special functional modules, should this be desirable. This circuitry is represented by the area to the right of the frame buffers in Figure 7. For neighborhood operations, an eight-bit image is selected by reading the contents of one of the two field buffers in the MPS. The image is transformed by a single-valued mapping through one of 32 program-selectable lookup-tables (pre-nop L.U.T.), and the pixels of the resulting image are passed to two neighborhood operators (NOP A, and NOP B), of which there are two kinds. The first type of neighborhood operator is an arithmetic convolution operation, while the second is a Boolean operation. For either operator, the neighborhood of operation is (at present) 3×3 pixels square, and the operation is accomplished in 200 nsec. Pixel neighborhoods are generated by passing the data stream through a 3-line buffer.

In the arithmetic case, the convolution operation uses arbitrary positive or negative eight-bit neighborhood weights, and maintains twelve-bit accuracy in its intermediate results. The final eight-bit arithmetic result busses is produced by non-biased rounding, from a 20-bit sum. This insures that no loss of precision occurs within a stage due to arithmetic underflow or overflow. The full eight bit precision of the input is thus maintained between stages throughout the machine. In the Boolean case, the neighborhood operation consists of arbitrary Boolean operations (a sum-of-products AND-OR array equivalent) between the set of all the pixels of the data neighborhood, and the set of all corresponding pixels of an arbitrarily specified comparison neighborhood. Any bit of either neighborhood may be independently defined as true, false (complemented), or "don't care". Each of the eight bit-planes forms an independent set of inputs, subject to independent neighborhood operations. As a result, eight independent one-bit results are obtained from a single pass of the data through the pipeline, yielding an orthogonal eight-bit Boolean vector as output.

Both neighborhood operators are applied independently and simultaneously. They operate on the same data stream, using neighborhood operations which may be different. Their outputs, or the contents of either of the field buffers, may be independently subjected to a second transformation by either of two programmable functions. The first of these is a lookup-table mapping function (TVF L.U.T. in Figure 7.) This transformation may be a function of one or two eight-bit arguments. If two arguments are used, they may be taken from homologous pixels of either of the NOP outputs or either of the field buffers. The lookup-table is program-selectable from a number of stored tables which depends on the number of arguments to the input. The other function, with inputs selectable from the same sources, is an ALU with two eight-bit inputs (ALU C.)

Finally, the contents of either of the buffers, the results of either of the neighborhood operators, and the results of either of the two functions of two arguments, may be sent to the three output pathways, or to the four wildcard busses, in any combination, by a crossbar switching network shown at the extreme right in Figure 7.

These basic features of the processing stages may be altered in operation may one of two special processing modes. The MIMD, or "region of interest" mode allows each MPS to switch between alternative operation sets on a pixel-by-pixel basis. In this mode, one image buffer of the stage contains a map of the operations to be performed on homologous pixels of the image buffer undergoing operations. In this mode, the contents of the operation-controlling buffer are treated as offsets into the on-board micro-operation store of the stage, and select the operative micro-operation code for each pixel-time. Each micro-operation word controls the routing of data-flow within the stage, the ALU functions, the identity of the look-up tables employed, and the selection of output paths. Potentially, up to 256 different alternative operation sets could be specified by the eight-bit contents of each pixel in the map. In practice, the number of alternative operation sets selectable during a field processing time will be limited by the amount of memory available within the stage to store them, which may be enlarged at will. The operation sets stored in the available memory may be changed arbitrarily between fields.

In "Pyramid Mode", PIPE allows the construction of multiresolution, "pyramid", sequences of images. The basic operations available in PIPE for constructing image pyramids sampling and pixel doubling. Sampling is used to reduce the resolution of an image, while doubling is used to increase the size of an image.

Both the sampling and doubling operations are performed by manipulating addressing rates within a stage. Sampling is achieved via the forward pathway by incrementing the destination image addresses half as fast as the source addresses. That is, on each row, the first pixel in the source image is written to the first pixel in the destination image. The second source pixel is also written to the first destination pixel. The address of the destination pixel is then incremented, and the procedure is repeated. The same process is used to sample into every other row in the destination image. The result is that the destination image is one quarter the resolution of the source image. Doubling is accomplished via the retrograde pathway by the inverse of the sampling process. That is, the addresses in the source image are now being incremented at half the rate of those in the destination image. For each row in the source (reduced-resolution) image, two identical rows are output in the destination image. For each pixel in each row of the source image, two identical pixels are stored in the destination image.

The simple operations of image sampling and pixel doubling are not of themselves very useful except for a narrow range of applications. However, they may be combined with the other operations in the MPS, so that a much broader class of operations becomes possible. Prior to sending the images over the forward and backward paths, they may pass through the operations of the output section of the MPS, and prior to storage, they may pass through the operations of the input section of the destination MPS. For example, the neighborhood operator can be used to smooth the image before sampling. By iterating the neighborhood operation prior to sampling, the effects of neighborhoods larger than three by three can be obtained, allowing, for example, the construction of "Gaussian" pyramids using the hierarchical discrete correlation procedure of Burt[3]. By passing an image through one neighborhood operator into the forward path, through the other into the recursive path, and by passing the retrograde path directly to the look-up table of the input section of the N-1st MPS, the pyramidal neighborhood operations of Tanimoto's Hierarchical Cellular Logic[4] may be realized.

Edge effects that arise when a neighborhood operator is applied are dealt with in the same manner for all resolutions of images, and for borders of MIMD operation regions as well as for frame borders. PIPE automatically provides the replication or zeroing of border pixels. If a neighborhood has a row or a column that lies outside the boundaries of the image (either beyond the image buffer itself or beyond the extent of a low-resolution image, or beyond the bound of a MIMD operation type), the non-existent pixels are zeroed or replaced by the pixels in the border row or column. For a three by three neighborhood, this is equivalent both to reflecting the image and to repeating the border pixels. This is achieved in the same way as the varying resolution images are constructed, i.e., by manipulating the address lines of the buffer.

Output Stage

The output stage performs a role at the end of the processing chain similar to that of the input stage at its beginning. The final MPS delivers its forward image output to either one of a pair of field buffers in the output stage, and can simultaneously read from the other buffer of the output stage. The data read from the output stage is used as the input to the retrograde path of the final MPS. Without interrupting the image-processing, either buffer of the output stage can be read from or written into by an external device, which is both the consumer of the processed forward data-flow and the supplier of data for the retrograde path.

Sequencer and Stage Control

Prior to run time, the pipeline modules must be set up with the individual operations to be carried out. The upper levels of the system must load each stage with appropriate instructions for all the processing steps to be employed. Each stage has storage for up to 256 128-bit micro-operation words, each of which can specify the entire set of operations for the stage. This store is loaded by the host or upper levels of the vision system. Each stage also has control circuitry which allows a micro-operation word to be selected from the stored set by external command. The selection command can come either from the host or from PIPE's Sequencer. In either case the selection can completely reconfigure the stage operation during the inter-field time by selecting a new stage operation set for the next field-time, or a group of stage operation sets from which members are to be selected on a pixel-by-pixel basis in MIMD mode during the next field-time.

Normally, the cycle-by-cycle selection of the operations stored in each stage will performed by the Sequencer module. When active, this unit issues an operation selection command to every stage of PIPE at the beginning of each field-time. The nature and order of these operation selection commands are determined by a program in the sequencer module which specifies the order of commands to the stages, including loops and branches, so that the coordination of stage operations to effect various PIPE algorithms is accomplished. This sequencer program is loaded from the host prior to run time. At any time, the host can override the sequencer program and intervene in the stage operation selection process directly. In operation, the upper levels of the system may instruct the sequencer to select a stage program, instruct it to branch in the specified sequence of operations, or permit it to follow the pre-set sequence of operations (which may contain branch points on repetition counts.)

Programming and running PIPE thus consists of specifying the operations to be performed by each stage, loading the corresponding operators, parameters, and tables into the stages, and then loading the sequence of operations for the stages into the sequencer module. For program development, the contents of any buffer and the output of any operator in the system can be displayed on a video monitor, while the sequencer is single-stepped.

INPUT/OUTPUT MODES

In keeping with its role as a real-time vision processor, PIPE has several rapid means for transferring information to and from the host device. The wildcard busses may be output directly, either as

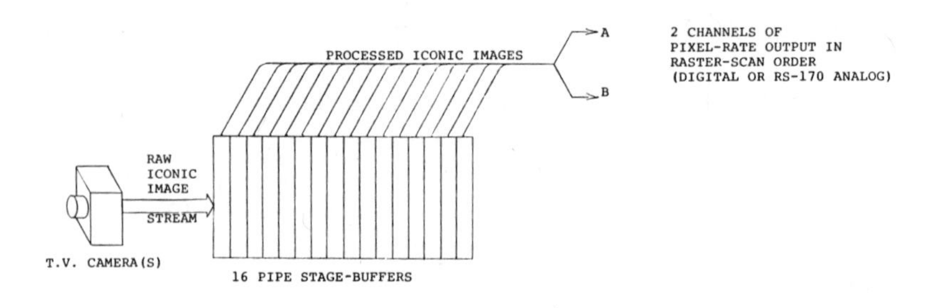

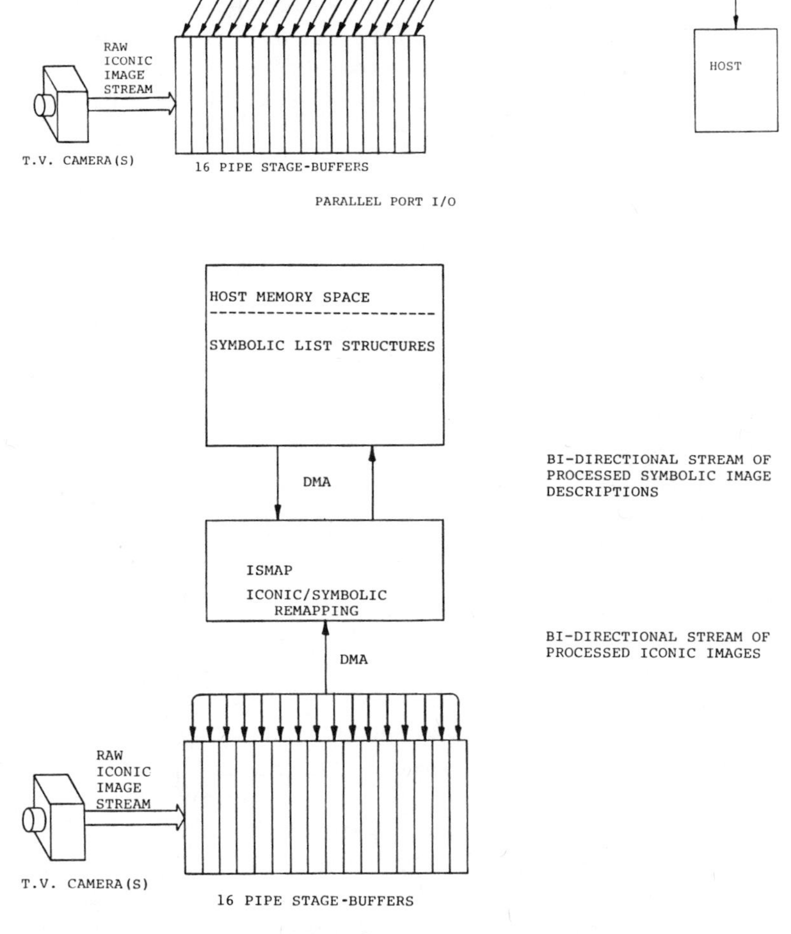

FIGURE 8 The I/O modes supported by PIPE. a.) streaming pixel I/O, b.) parallel port random access, c.) bi-directional ISMAP operation.

digitial streams or after conversion to analog signals. The analog outputs provide for interface to monitors or other display devices, while the digitial outputs provide streaming I/O at video rates to any specialized device which can accept this rate of input. This streaming I/O mode is represented in figure 8-a. Figure 8-b presents a parallel port I/O scheme which allows the host device access to any of the PIPE image buffers in a random access fashion. These parallel port transfers can also function in auto-increment and DMA modes. The most sophisticated method of transferring image information to the host uses the ISMAP device to transfer symbolic descriptions of the image to the host's address space. This transfer may be bidirectional, and can take any PIPE buffer as a source or destination. This I/O technique is illustrated in Figure 8-c.

ISMAP

The associated device, ISMAP, will map processed iconic images into symbolic (property-indexed) structures. ISMAP describes the processed images produced by PIPE by mapping the iconic image of Boolean vectors produced by PIPE into ordered list structures in the address space of processors in the upper levels of the system, in real time. A major function of ISMAP is to map PIPE's image addresses into a space ordered by symbolic picture feature descriptors. This saves processors in the upper levels of the system the necessity of scanning the processed image to find locations of items of interest. Physically, ISMAP resides in the PIPE backplane as an integrated part of the system.

In each frame time, ISMAP scans the entire image of iconically-ordered feature codes produced by PIPE, and prepares three organized descriptions of the feature set of the image in a symbolically-ordered space in the address space of the host. The three descriptions of the image's feature content produced by ISMAP are: 1) a histogram of feature types, organized by feature type, 2) a cumulative histogram of feature types organized by feature type, and 3) lists, organized by feature type, of the image-coordinates of every feature in the image. The values in the cumulative histogram serve as pointers to the heads of the lists of image locations. For example, if the host needs the locations of all edges with a certain angle of inclination, it can find the beginning and the length of a compact list of those locations, in its own address space, simply by looking at the cumulative histogram entry for the desired edge direction. It may first use the histogram to determine what edge directions are prominent in the image, or perform any of a variety of other algorithms on this information without the burden of examining the image-space in order to count or locate features of interest.

ISMAP may also run in reverse, creating images in host buffers from feature lists created or selected by the host. This permits ISMAP to be the first level of the upper system which can generate hypothesis images for PIPE, which allows the PIPE/ISMAP combination to perform operations such as Hough transforms easily.

CONMAP

ISMAP is one of two PIPE auxiliary devices. Another, CONMAP, which is currently under construction, stands at the front of PIPE as a geometric pre-processor for input images. The CONMAP device accepts iconic images and remaps them into topologically identical, but geometrically different, iconic images prior to their acquisition by PIPE. In its final form, CONMAP will implement fully general conformal mappings on images. There are a variety of uses for such transforms. Log-polar mappings have been suggested by Weiman and Chaikin[5], as a means of converting complex image properties, such as rotation and scale, into simple translations. Jain[6] has employed a similar transformation for utilizing the effects of camera motion in image understanding. Other applications include simulation of lens geometries which permit non-uniform resolution (e.g. a high-resolution 'fovea' with lower-resolution periphery.) CONMAP will transform images at field rate, and its operation will be transparent to PIPE except for a one-field delay.

PIPE's SYSTEM SOFTWARE

PIPE is provided with a software toolkit which includes a number of highly-developed programming systems. A simulator exists, although it is not used when a PIPE machine is available, since the other software development tools permit interactive test and debugging of algorithms in real-time on the machine. All PIPE system software is graphics-based and runs in color-coded formats on the screens of the host.

At the bottom level, a software "keyboard" permits interactively displaying and setting any bit, byte, or word in the machine's micro-coded operation stores. This can be done while PIPE is running algorithms, and the effects observed. This program can also save code to or load code from disk files.

Microcoded instruction lists for the operation store can be produced and stored in disk files by a graphics-based assembler. This program presents diagrammatic icons portraying the logical functions of the processing stages on the screen, and interactively fills in data-flow paths and operations as the user specifies them. The program has an expert-system knowledge of PIPE. It will only inquire about options as they become necessary based on user choices, and it will reject contradictory or illegal choices.

Figure 9 demonstrates a PIPE programming diagram. Spatial progression of stages is from left to right. Temporal progression of machine cycles is from top to bottom. A single row indicates the state of the machine in a single field time; a single column indicates the successive states of a single stage. Diagrams such as this illustrate the space-time systolic nature of PIPE programs, and can be transferred directly to the graphics screens of the system software tools for program generation.

Another graphics-oriented program, similar to an assembler, generates programs for the global sequencer which, in turn, controls selection of words in the micro-operation stores of all PIPE's stages on a cycle-by-cycle basis. This program generates disk files which can be loaded into PIPE's system controller.

PIPE employs many tables of great complexity. A table-generating program is available to simplify their production. For example,

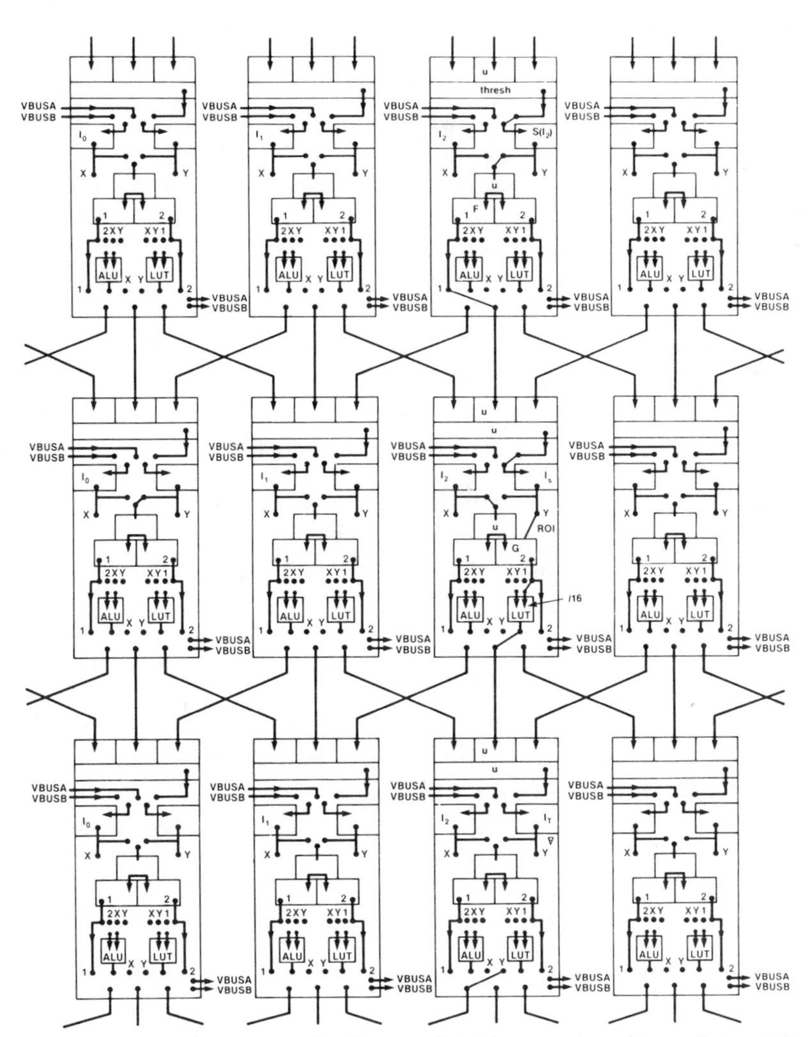

FIGURE 9 A PIPE programming diagram. Spatial progression of stages is from left to right. Temporal progression of machine cycles is from top to bottom. A single row indicates the state of the machine in a single field time; a single column indicates the successive states of a single stage.

extremely complex tables are required for the bit-slice arithmetic in the convolvers. The table entries have no simple relationship to the weights of the neighborhood mask. The table-generating program will present a graphic representation of the neighborhood, query the user for mask values, and prepare the required tables and save them in disk files for later loading into PIPE. A variety of other table types are also handled. A related program finds the optimal sequence of 3×3 kernels for the production of any $N \times N$ convolution.

At the next level, PIPE is programmable in PIPE Command Language (PCL). In PCL, it is possible to refer to stages, storage areas, and operational units of the machine, to command their states, and to load them from buffers in the host.

A graphics-oriented interactive monitor program permits real-time manipulation of PIPE with PCL. This functions much like a low-level monitor program on a traditional computer, except that the command set is the full set of a powerful macro-assembly language for the machine. The monitor is useful principally in debugging programs that have been generated by other means, but it is possible to build short test programs with it.

The principal means of generating PCL programs is by means of a low-level compiler which accepts commands in PIPE Intermediate Format (PIF). PIF is a higher-level language which permits the programmer to command successive states of machine operations, to refer to named disk files, and to associate the files with the machine entities. Such files may be the micro-operation lists, sequencer programs, and tables generated by auxiliary programs. These files represent operations of broad general utility, and form a library upon which PIF programmers can draw.

This compiler can operate either from PIF programs stored in disk files, or interactively with the user as he issues PIF commands. Most of the existing applications software for PIPE has been written in PIF. It is intended as the intermediate format which will be generated by compilers for higher-level image-processing languages now under construction.

Software tools already available for PIPE would permit a competent PIPE programmer to write all PIPE software required for most applications at a realistic level and with reasonable efficiency, although the task would be made easier if higher-level software tools currently being produced were available now. A growing body of basic application routines, for feature detection, image enhancement, and similar elementary functions already exists. About one day's time is sufficient for a competent programmer to write such a routine. Experience suggests that a doctoral-level computer scientist can become an expert PIPE programmer in less than one month's time.

DISCUSSION

PIPE is designed as a front-end processor for low-level iconic-to-iconic image processing. It is intended to perform transformations on images to extract features similar to those in the primal sketch of Marr.[7] These features make intensity changes and local geometric relations explicit in images, while maintaining the spatial representation. In this, PIPE differs from many processors designed for image-processing. These other processors are usually designed to perform both local and global image-processing tasks, often in an interactive environment.

A recent survey by Reeves[8] divides image-processing tasks into two classes. Low level image processing usually modifies parts of images, but maintains the image array. Higher level processing, however, works on symbolic representations of the contents of images. Low level processing has usually given rise to architectures based on single instruction stream, multiple data stream (SIMD) structures. The higher level functions are usually carried out using processors based on multiple instruction stream, multiple data stream (MIMD) structures. The design of PIPE allows it to act as a SIMD pipeline, or as a (restricted) MIMD pipeline. The MIMD mode is entered whenever the region-of-interest operators are used. The limitations on these operators are that there are at most 256 different operators available per stage, and that using the region of interest generally precludes using some other operators, such as the functions of two arguments. Using the retrograde pathway to insert expectations from the host into the image analysis process also blurs the distinction between high level and low level processes.

Some general features of PIPE's architecture, multiple pipelined planes of processing, and the concept of the iconic-to-symbolic remapping performed by ISMAP, are drawn in spirit from a more elaborate machine proposed but never constructed by McCormick,

Kent, and Dyer.[1,2] That device was in turn inspired by certain observations on the nature of processing in the visual cortex. Although PIPE is in some respects a simpler device, it also carries that analogy further with the implementation of the backward pathway which emulates the connectivity in the biological vision system, where, with the exception of the final link between the retina and the lateral geniculate nucleus, there are in fact more fibers descending from higher to lower levels of processing than *vice versa*. The arrangement of interactions among the three pathways, with combination of images into stage buffers and separate operations available into emerging pathways also is drawn from this model.

CONCLUSIONS

This paper has described a new image pre-processor, consisting of a sequence of identical stages, each of which can perform a number of point and neighborhood operations. An important feature of the processor is the provision of forward, recursive, and backward paths to allow image data to participate in temporal as well as spatial neighborhood operations. The backward pathway also allows expectations or image models to be inserted into the system by the host, and to participate in the processing in the same way as images acquired from the input device. The region-of-interest operator is also a powerful, and unique, feature of PIPE, allowing the results of feature-extraction processes to guide further image analysis. PIPE also provides a multi-resolution capability, enabling global events to be made local. This is important in a machine that has only local operators. Much research needs to be done to explore the capabilities of the device but early experiments indicate that the system will have a wide range of applications in low-level real-time image processing.

REFERENCES

1. B. McCormick, E.W. Kent, and C. Dyer, A Visual Analyzer for Real-Time Interpretation of Time-Varying Imagery., in: *Multicomputers and Image Processing 3.*, K. Preston and L. Uhr (Eds.), Academic Press, 1982.
2. B. McCormick, E.W. Kent, and C. Dyer, A Cognitive Architecture for Computer Vision. in: *Fifth Generation Computer Systems.*, T. Moto-Oka (Ed.), North Holland, 1982.
3. P.J. Burt, Fast hierarchical correlations with Gaussian-like kernels. Proc. Fifth International Joint Conference on Pattern Recognition, Miami, Florida, 1980.
4. S.L. Tanimoto, A Hierarchical Cellular Logic for Pyramid Computers. J. Parallel and Distributed Computing, 1, 105–132, 1984.
5. C. Weiman, and G. Chaikin, Logarithmic Spiral Grids for Image Processing and Display., Computer Graphics and Image Processing, *11*, 197–226, 1979.
6. R. Jain, Segmentation of Frame Sequence Obtained by a Moving Observer., General Motors Research Publications: GMR-4247, 1983.
7. D. Marr, Early processing of visual information. Phil. Trans. Royal Society B.275, 1976.
8. A.P. Reeves, Parallel computer architectures for image processing. Computer Vision, Graphics, and Image Processing 25, 1984, 68–88.

Chapter 5

ARCHITECTURES FOR ROBOT VISION*

T.N. Mudge
and
T.S. Abdel-Rahman

Department of Electrical Engineering and Computer Science
University of Michigan
Ann Arbor, MI 48109

INTRODUCTION

Interest in the area of computer processing of visual data has increased considerably over the past decade. This has led to an increasing number of applications for computers in that area. The applications vary from Robot Vision in industrial environments to Tomography in medical environments. It is safe to predict that the range of applications will continue to expand in the near future to cover many other application areas and give a wider scope to already existing applications. A partial list of applications include the following [1]:

- Automation of Industrial Processes.
  - Object acquisition together with a robot arm ('bin of parts problem').

* This work was supported in part by the Materials Laboratory, Air Force Wright Aeronautical Laboratories, Aeronautical Systems Division (AFSC), United States Air Force, Wright-Patterson AFB, Ohio 45433-6503.

- — Automatic tool guidance.
- — Visual feedback for automatic assembly and repair.
- — VLSI (Very Large Scale Intergration) circuit inspection and checking.
- — Inspection of printed circuit boards.
- — Screening and inspecting plant samples.
- — Inspection of castings for impurities and fractions.

- Medical Applications.
  - — Tomography.
  - — Enhancement and analysis of X-ray images.
  - — Blood cell recognition and counting.
  - — Tissue and chromosome analysis.
  - — Aiding the blind.
- Remote Sensing.
  - — Cartography.
  - — Monitoring traffic.
  - — Exploration of remote and hostile areas.
- Military Applications.
  - — Tracking moving objects.
  - — Automatic navigation.
  - — Target acquisition and range finding.

There are several disciplines that are concerned with the processing of visual data or images. The primary ones are: *Image Processing*, which is concerned with the transmission, storage, enhancement and restoration of images; *Pattern Recognition*, which is concerned with classifying regions of the input image as one of a (usually small) set of possibilities; and *Image Understanding*, which is aimed at the construction of a rich set of descriptors of images to facilitate the interpretation and the understanding of those images [1]. Secondary disciplines include: *Computer Graphics* and *Computer Aided Design*. Systems that process images benefit from the above disciplines to varying degrees depending on the specific nature of the application at hand.

The objective of this chapter is to study special purpose architectures for Robot Vision. This will be accomplished in two steps. First, algorithms for Robot Vision are discussed and a set of benchmarks of typical algorithms for Robot Vision is presented. Second, various classes of special purpose architectures for Robot Vision are discussed. The set of benchmarks is then used to evaluate the performance of the various architectures and illustrate their basic characteristics.

The remainder of the chapter is organized as five sections: the rationale behind special architectures for Robot Vision; basic terminology that is used throughout this chapter; algorithms for Robot Vision including the set of benchmark algorithms; special computer architectures for Robot Vision and their performance on the benchmarks; finally, conclusions and final remarks.

RATIONAL FOR SPECIAL ARCHITECTURES

Robot Vision applications require very high computational power that is beyond the capabilities of modern conventional computers. This is principally due to the massive amounts of data that has to be processed in such applications. Images, typically represented as two dimensional arrays of M^2 pixels (picture elements), may have to be processed at the rate of 30 images per second to meet real-time video rate requirements. Typical values of M range from 256 to 4096. In some applications, however, M can be as large as 10^4. Each pixel in the image can have from 1 bit to 24 bits of data representing a gray level value or a color intensity.

To illustrate the computational effort needed to process images, Table 1 shows the processing time for a $512 \times 512 \times 1$ byte image on a conventional computer such as the VAX 11/780 supermini, which is capable of 1 MIPS (Million Instructions Per Second). The first column shows the number of operations performed on each pixel in the image. The second column shows the approximate number of instructions needed to process the image. An average of four instructions per operation is assumed. The third column gives the execution time for the image. The numbers indicate that a conventional computer would fall well behind the required computational effort. Larger images require even longer processing times. Furthermore, 30 images per second must be processed if video rates are required.

Hence, it becomes necessary to employ special purpose machines

Ops per pixel	Total No. of instrs.	Time for a VAX
10^2	10^8	100 sec
10^3	10^9	17 min
10^4	10^{10}	2.78 hr

Table 1 Processing times for a 512 × 512 × 1 byte image on a VAX 11/780

that have sufficient processing capabilities to provide the necessary speedup over conventional computers. This can be achieved through three means:

1. Using a large number of processors working in parallel. This is referred to as *Parallel Processing*.
2. Using faster computer hardware technology.
3. Using better programming through optimizing compilers and new algorithms.

Parallel processing is the basic means for providing computational power for Robot Vision. Newer computer hardware technologies, while providing high levels of integration that roughly quadruple every two years, have only been able to increase speed by a factor of 10 in the past decade [2]. Better programming cannot provide a substantial speedup in Robot Vision applications due to the simple but repetitive nature of the operations involved.

Furthermore, the nature of image data and the nature of Robot Vision algorithms (as will be clear later) make parallel processing very attractive. Many Robot Vision algorithms involve operations that transform the value of each pixel in the image based on the values of surrounding pixels. This locality in operations suggests that image data be decomposed into a large number of subsets that can be processed in parallel. Hence, it is not surprising that an architectural characteristic of most special processors for Robot Vision is the employment of parallel processing to achieve the necessary computational power.

TERMINOLOGY

An image, a frame or a picture, P, is a two dimensional array of pixels, $P(i,j)$, $i,j = 1,\ldots, M$. The term $P(i,j)$ can stand for the name of the pixel at (i,j) or the gray level value at (i,j). The gray level value is typically an integer in the range $0,\ldots,255$. $P(0,0)$ is located at the top left corner of the array. The index i runs in the negative y-direction. The index j runs in the x-direction. This is shown in Figure 1.

A neighborhood of a pixel $P(i,j)$, denoted by $N(i,j)$, is a set of contiguous pixels that include $P(i,j)$. Some of the common neighborhoods are:

- $m \times m$ neighborhood, (m odd). Defined as

$$N(i,j) = \bigcup_{\alpha,\beta} P(i + \alpha, j + \beta) \qquad \alpha,\beta = -\left\lfloor \frac{m}{2} \right\rfloor, \ldots, \left\lfloor \frac{m}{2} \right\rfloor.$$

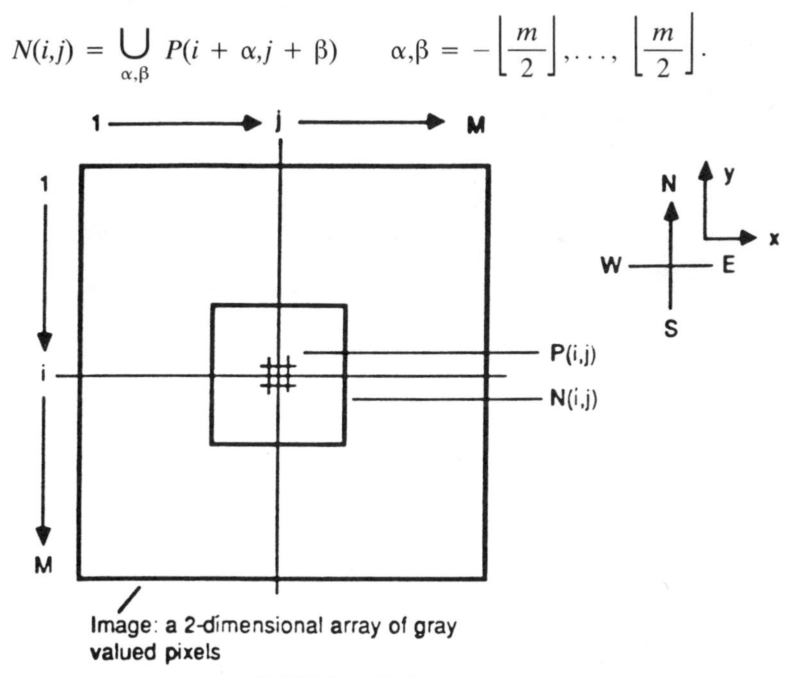

FIGURE 1 An image $P(i,j)$.

FIGURE 2 The heap-of-parts image.

The 3×3 neighborhood is common to many Robot Vision algorithms.

- $m \times 1$ neighborhood, defined as

$$N(i,j) = \bigcup_{\alpha} P(i + \alpha, j) \qquad \alpha = -\left\lfloor \frac{m}{2} \right\rfloor, \ldots, \left\lfloor \frac{m}{2} \right\rfloor.$$

- $1 \times m$ neighborhood, defined as

$$N(i,j) = \bigcup_{\beta} P(i, j + \beta) \qquad \beta = -\left\lfloor \frac{m}{2} \right\rfloor, \ldots, \left\lfloor \frac{m}{2} \right\rfloor.$$

Throughout this chapter, the image shown in Figure 2 will be used to illustrate Robot Vision algorithms. The image contains a heap of industrial parts.

ALGORITHMS FOR ROBOT VISION

The processing of images for Robot Vision proceeds in the three major stages. In the first stage, referred to as the *low level processing* stage, the image is enhanced and features in the image are detected. Algorithms that perform these functions are referred to as *low level algorithms*. Low level processing draws most of its techniques from the field of image processing. Hence, the low level processing stage may be considered the stage of image processing. In the second stage, referred to as the *intermediate level processing* stage, the features detected in the first stage are extracted from the image. Algorithms that perform such functions are referred to *intermediate level algorithms*. In the third and final stage, referred to as the *high level processing* stage, the extracted features from the image in the second stage are classified and analyzed. Algorithms that perform these operations are referred to as *high level algorithms*. High level processing draws most of its techniques from the fields of pattern recognition and image understanding. Hence, the high level processing stage may be considered the stage of image analysis and understanding.

The above mentioned stages of processing will be referred to as the *Robot Vision System (RVS)*. The RVS is depicted in Figure 3.

In this section, the basic characteristics of each processing stage and its algorithms are discussed. Benchmarks of typical algorithms from each stage are also presented.

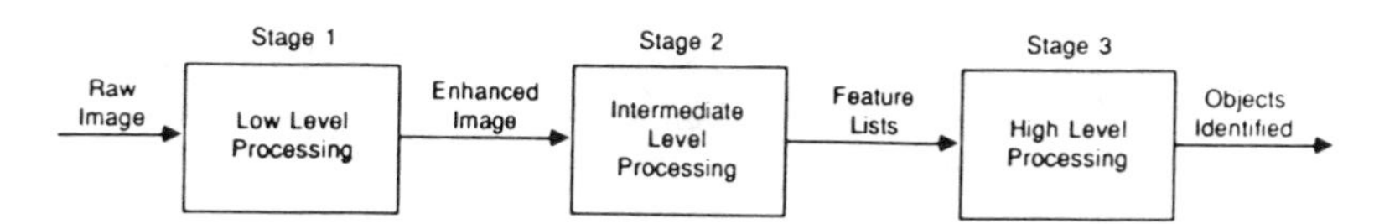

FIGURE 3 The Robot Vision system.

Low Level Processing

Low level processing generally involves enhancement, restoration, noise removal and feature detection operations. This stage of processing is characterized by:

1. The image is represented as a two dimensional array of pixels. Each pixel represents the gray level value at its coordinates. This representation of the image, which is referred to as a *low level representation*, can be characterized in the following manner:
   (a) The spatial information regarding pixel location is implicit in the representation of the array. That is, no explicit address information is stored.
   (b) The classification of pixels is explicit in the representation through the values of the pixels in the array.
   (c) The features in the image are implicit in the representation through the relationships among the pixels.
2. A set of deterministic operations is applied to each pixel in the image. As a consequence, equal size areas of the image take equal amounts of processing times. Furthermore, because of the deterministic nature of the operations, no data dependent decisions are made during run-time. These operations basically are of four types[3]:
   (a) *Input/Output Operations*: for human interaction and image storage/retrieval.
   (b) *Context-Free Operations*: point-wise operations on single or multiple images. Examples include histogram generation and general tonal mapping.
   (c) *Context-Dependent Operations*: the value of a pixel is modified based on the values of the pixels in its context or neighborhood. Hence, these operations are also referred to as neighborhood operations. Context dependent operations form the majority of low level operations. Examples include: edge and line detection, noise removal and filtering.
   (d) *Global Transformation Operations*: the image is transformed by Fourier, Cosine, Hadamard and similar transforms.

The following are the set of low level algorithms used in our benchmark. They illustrate the general characteristics described above for low level algorithms.

Convolving with an FIR function

In this algorithm, the input image is convolved with a Finite Impulse response (FIR) function to give the output image. The FIR function is a matrix $K(\alpha, \beta)$ of constant coefficients of dimensions $m \times m$. This FIR matrix is also referred to as the *kernel*. The kernel is moved across the image in one pixel steps to implement the convolution function. In each step, the pixel $P(i,j)$ coinciding with the center of the kernel is replaced by $Q(i,j)$, where

$$Q(i,j) = \sum_{\alpha}\sum_{\beta} P(i + \alpha, j + \beta)K(\alpha, \beta).$$

Convolving the image with a FIR function falls into the type of context dependent operations. The kernel $K(i,j)$ can be viewed to coincide with the $m \times m$ neighborhood for the pixel $P(i,j)$. The value of $P(i,j)$ is then replaced by a new value, $Q(i,j)$, which is a function, linear in this case, of the pixels in the neighborhood. This is shown in Figure 4.

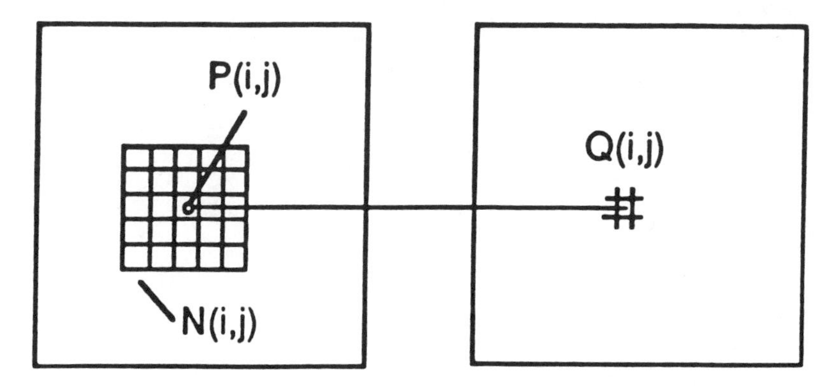

FIGURE 4 Convolving with the FIR function.

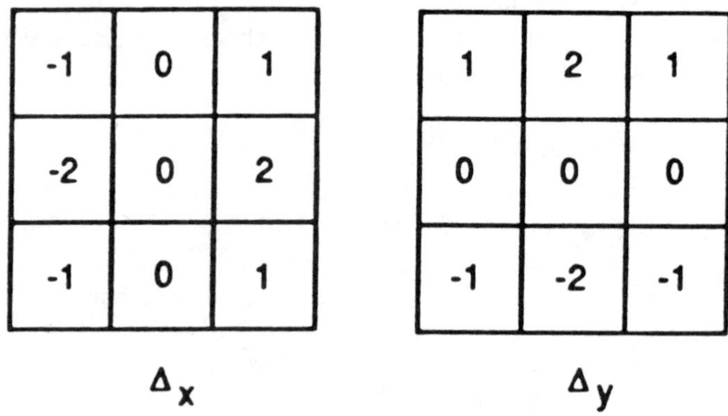

FIGURE 5 The Sobel edge detector kernels.

A frequently used example of an algorithm involving the convolution of the image with FIR filters is the use of the Sobel operators for edge detection[4]. The image is convolved with each of the two FIR kernels Δ_x and Δ_y shown in Figure. 5. The result of the convolution are the two images e_x and e_y where:

$e_x(i,j)$ is an $M \times M$ array of x-direction edge (gradient) strengths.
$e_y(i,j)$ is an $M \times M$ array of y-direction edge (gradient) strengths.

These two images are then combined to form a combined edge strength array, E, and an edge direction array, θ, where

$$E(i,j) = \sqrt{e_x^2 + e_y^2},$$

and

$$\theta(i,j) = arctan\left(\frac{e_y}{e_x}\right) - \frac{\pi}{2}.$$

We assume that the bright side is to the left of the edge when pointing in the direction of the edge. A numerical example that illustrates the Sobel edge detection operators is shown in Figure 6. The heap-of-parts image is shown in Figure 7 after the application of the Sobel operators.

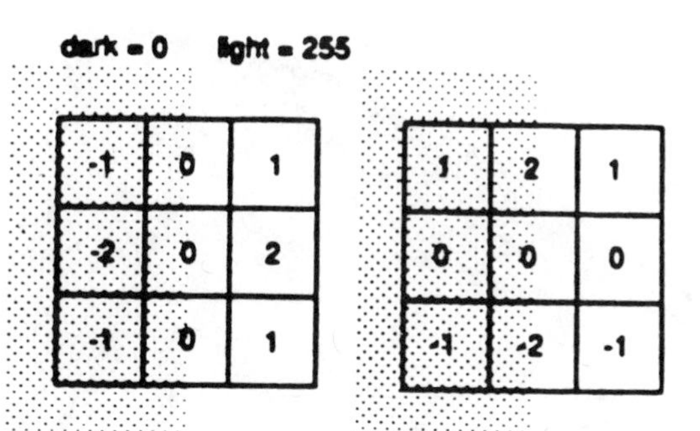

$$e_x = \frac{1020}{4} = 255 \qquad e_y = 0$$

$$E = 255 \qquad \Theta = \arctan(0) - \frac{\Pi}{2} = -\frac{\Pi}{2}$$

-1	0	1
-2	0	2
-1	0	1

1	2	1
0	0	0
-1	-2	-1

$$e_x = 191.25 \qquad e_y = -191.25$$

$$E = 270.47 \qquad \Theta = \arctan\left(\frac{-1}{1}\right) - \frac{\Pi}{2} = -\frac{3\Pi}{4}$$

(E Would be 255 if $\frac{1}{\sqrt{2}}$, 1, $\frac{1}{\sqrt{2}}$ weights were used.)

FIGURE 6 A numerical example of the Sobel operators.

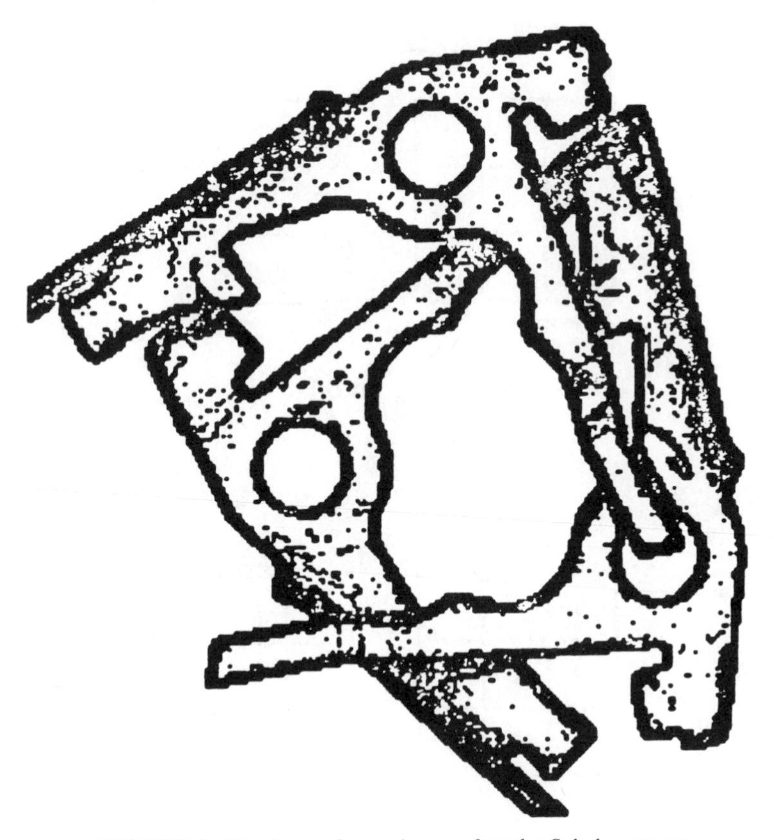

FIGURE 7 The heap-of-parts image after the Sobel operators.

Convolve or Transform

When the convolution kernel $K(i,j)$ becomes large enough, it is more computationally efficient to perform the convolution in the frequency domain by utilizing the convolution theorem[5,23]:

$$K * P = \mathcal{F}^{-1}(\mathcal{F}(K) \times \mathcal{F}(P)).$$

Where K and P are the kernel and the image to be convolved respectively, * denotes the convolution operation, $\mathcal{F}$ is the Discrete Fourier Transform (DFT) and $\mathcal{F}^{-1}$ is the inverse DFT.

In the spatial domain, the time to convolve the $M \times M$ image P with the $m \times m$ kernel K, in units of the time for an addition, is given by

$$T_s = m^2 M^2 (1 + \mu),$$

where μ is the time for a multiplication in units of the time for an addition (we have ignored boundary effects).

In the frequency domain, it is necessary to extend both the image and the kernel with 0's so they have the same period L, i.e., they are both $L \times L$ matrices. To avoid wraparound errors, the value of L must be chosen such that the condition $L \geq M + m - 2$ is satisfied[5].

The DFT of an $L \times L$ matrix Q, which is given by

$$\mathcal{F}(P(k,l)) = \sum_{i=0}^{L-1} \sum_{j=0}^{L-1} P(i,j) W_L^{ki} W_L^{lj},$$

where $W_L = exp\left(\frac{-2\pi\sqrt{-1}}{L}\right)$, can be split into two Fast Fourier Transforms (FFT's) and performed in $L^2\log_2(L)$ complex multiplications and $2L^2\log_2(L)$ complex additions. The same count of operations is incurred when the inverse DFT is performed.

Therefore, the application of the convolution theorem can be performed in $3L^2\log_2(L) + L^2$ complex multiplications and $6L^2\log_2(L)$ complex additions. Assuming that a complex addition is two normal additions and a complex multiplication is four normal multiplications plus two normal additions, then, in units of time for an addition, the time to convolve the image in the frequency domain is given by

$$T_f = 6(3 + 2\mu)L^2\log_2(L) + 2(1 + 2\mu)L^2.$$

Then direct convolution should be used only if

$$T_s < T_f.$$

Assuming $\mu \gg 1$ and $M \gg m$ then the above condition becomes

$$m^2 > 12\log_2(M) + 4,$$

which requires that m be about 9 for a 512×512 image.

The above results hold only for the serial implementation of convolution. In the case of parallel implementation, data movement time, if necessary, must be taken into consideration as well.

Non-Maximal Suppression

The non-maximal suppression algorithm is used to thin the thick edges obtained by edge operators such as the Sobel edge operator to exactly one pixel wide edges. It performs that function by suppressing (i.e. eliminating) locally non-maximal strength edge points in the direction normal to the edge direction[6].

The algorithm is applied to each pixel $P(i,j)$ in the input image and needs edge strengths at 3 points in the image: the pixel $P(i,j)$ and two points that lie on the normal to the edge direction at the pixel $P(i,j)$. This is depicted in Figure 8. The edge vector at $P(i,j)$ which has magnitude $E(i,j)$, the edge strength, and slope $\theta(i,j)$, the edge direction, is shown as the vector $\vec{E}$. The two points P_s and P_n are obtained by extending the normal to the edge direction

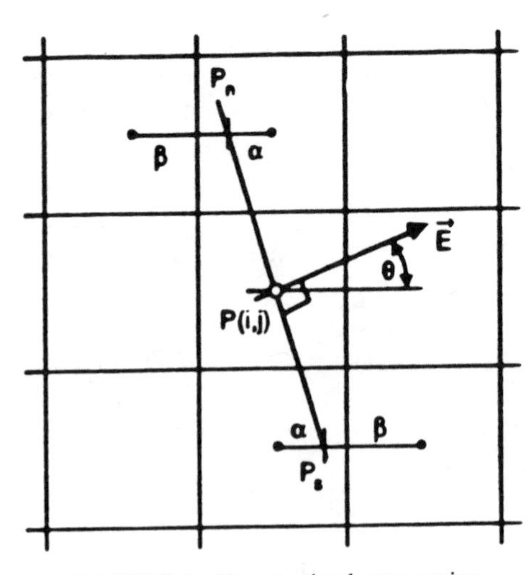

FIGURE 8 Non-maximal suppression.

as shown in the figure. As the edge strengths are available only at discrete intervals, it becomes necessary to interpolate for the edge strengths at points P_s and P_n as follows:

$$E_n = \beta E(i - 1,j) + \alpha E(i - 1,j - 1)$$

and

$$E_s = \beta E(i + 1,j + 1) + \alpha E(i + 1,j),$$

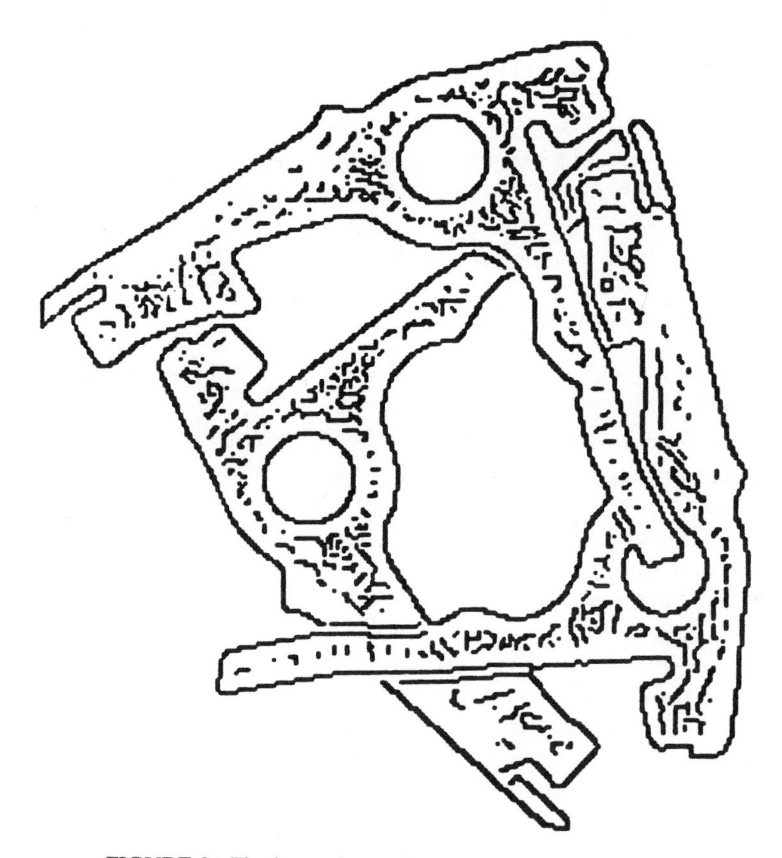

FIGURE 9 The heap-of-parts image after low level operations.

where E_s and E_n are the edge strengths at points P_s and P_n respectively, $\alpha = \left|\frac{e_x}{e_y}\right|$ and $\beta = 1 - \left|\frac{e_x}{e_y}\right|$.

The suppression of the non-maximal point is then implemented as

$$\text{If } E(i,j) < E_n \text{ or } E(i,j) < E_s \text{ then eliminate } E(i,j).$$

The heap-of-parts image is shown in Figure 9 after edge detection and the non-maximal suppression.

Intermediate Level Processing

The stage of intermediate level processing involves the extraction of features from the enhanced image produced by low level processing. The extracted features are then re-represented in a more convenient data structure to facilitate their high level processing. Hence, this stage is basically a transducer stage between the low level and the high level processing stages. This stage of processing has the following general characteristics:

1. The representation of the image in the input to the processing is a low level representation.
2. The representation of the image in the output is a set of attributes that explicitly describe the features in the image. That is, the low level representation of the image at the input, in which the features in the image are not explicit, is transformed into another representation in which features are explicitly and, hence, more conveniently represented. In general, it is difficult to characterize the data structure representation at the output of intermediate level processing any further. This can be attributed to the fact that efficient and convenient representations of different features require different data structures. Furthermore, different algorithms may elect to represent the same feature in different ways depending on the use of the feature in the algorithm. However, the following can be said about these representations:
   (a) The spatial information regarding the location of the pixels is explicit in the representation.

(b) The representation of a pixel value is generally restricted to a small set of values indicating the membership of the pixel in a feature set.

3. The set of operations applied to the pixels of the image generally depends on the value of the pixel and/or its context. As a consequence, equal size areas of the image take different amounts of processing times.

The following algorithms form the benchmark for intermediate level algorithms.

Edge Following Algorithm

Although edge detection and non-maximal suppression yield one pixel wide edges, these edges are still not related or organized in any way that facilitates their further processing. Furthermore, there is a large number of pixels that do not represent any edge points and, hence, are of no further interest. The edge following algorithm can be used to extract the edge points and organize them in the form of connected boundary segments. It also discards pixels that are not part of these boundary segments.

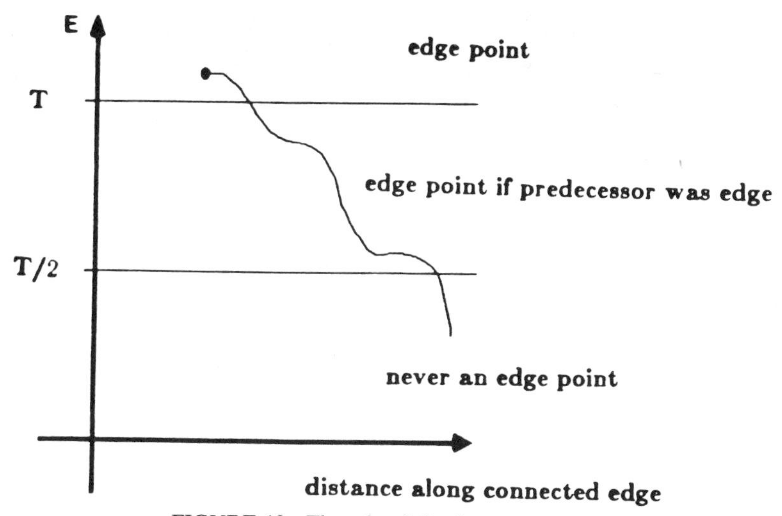

FIGURE 10 The edge following algorithm.

The edge following algorithm works as follows[7]. Starting at any edge pixel in the image, the algorithm traces the boundary by moving from one edge pixel to its neighboring one. If at any pixel the edge strength is greater than T, where T is a predefined threshold, then the pixel belongs to the boundary segment. If the edge strength at the pixel falls between $\frac{T}{2}$ and T, then the pixel belongs to the boundary only if its predecessor pixel is already on the boundary. Finally, if the edge strength at the pixel is less than $\frac{T}{2}$ then that pixel is discarded. At this point the boundary segment terminates and all edge pixels that belong this boundary segment are removed from the image and are represented by a linked list. Each link in the list has information regarding the edge strength, edge

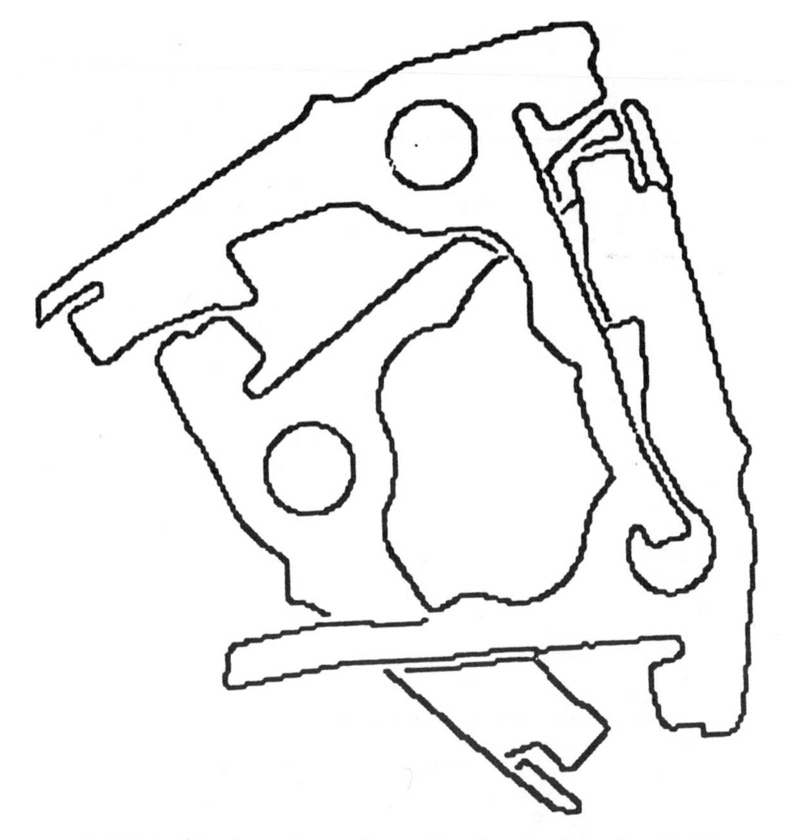

FIGURE 11 The heap-of-parts image after the edge following algorithm.

direction and coordinates of one pixel in the boundary segment. The edge following process is depicted in Figure 10. The above process then repeats starting at a new edge point in the image. The edge following algorithm terminates when there are no more edge points in the image. The output of the algorithm is, hence, a set of linked lists each describing a boundary segment in the image.

The edge following algorithm illustrates the basic characteristics of intermediate level algorithms. The features, in this case edge pixels, are extracted from the image and re-represented along with explicit address information in another data structure, in this case a linked list. Furthermore, the processing of equal area images require, in general, different amounts of processing times as the images contain, in general, unequal counts of edge pixels.

The heap-of-parts image is shown in Figure 11 after the edge following algorithm.

The Hough Transform

The Hough transform algorithm [7,25] can be used to identify edge pixels that form straight lines in the image. It does so by considering all possible lines at once and then rating each on how well it explains the the original image data.

The basic concept behind the Hough transform is the relation between points in the *image space* and lines in the *parameter space*. This relation is illustrated in Figure 12. For a point (X_1, Y_1) in the image space, the set of all possible lines that can pass through that point must satisfy the equation $Y_1 = mX_1 + c$. If (X_1, Y_1) are considered to be fixed, allowing m, c to be variables, then this point corresponds to the line $c = -X_1m + Y_1$ in the c-m space or the parameter space. Hence, a point (X, Y) in the image space corresponds to a line in the parameter space governed by the equation $Y = mX + c$. Then if the two points (X_1,Y_1) and (X_2,Y_2) are connected by the line AB, which is described by the equation $Y = m_1X + c_1$, then there must exist two lines in the parameter space corresponding to these two points which intersect in the point (m_1,c_1). By using the same argument, all points that are on the line AB must correspond to lines in the parameter space that intersect at the point (m_1,c_1).

The relation between the image space and the parameter space can be used to implement the Hough transform as follows. The

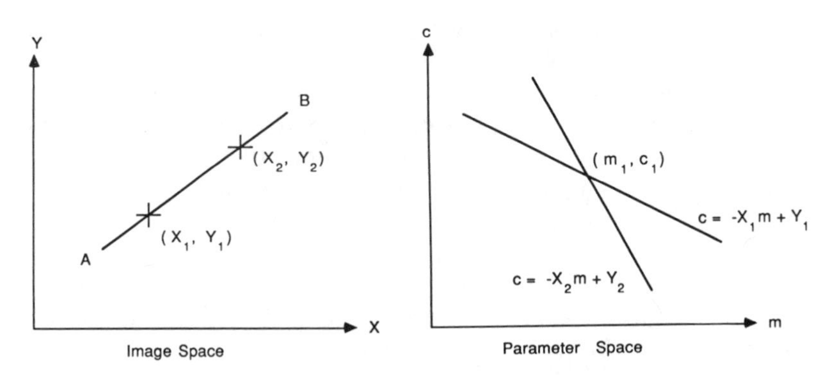

FIGURE 12 The relation between image space and parameter space.

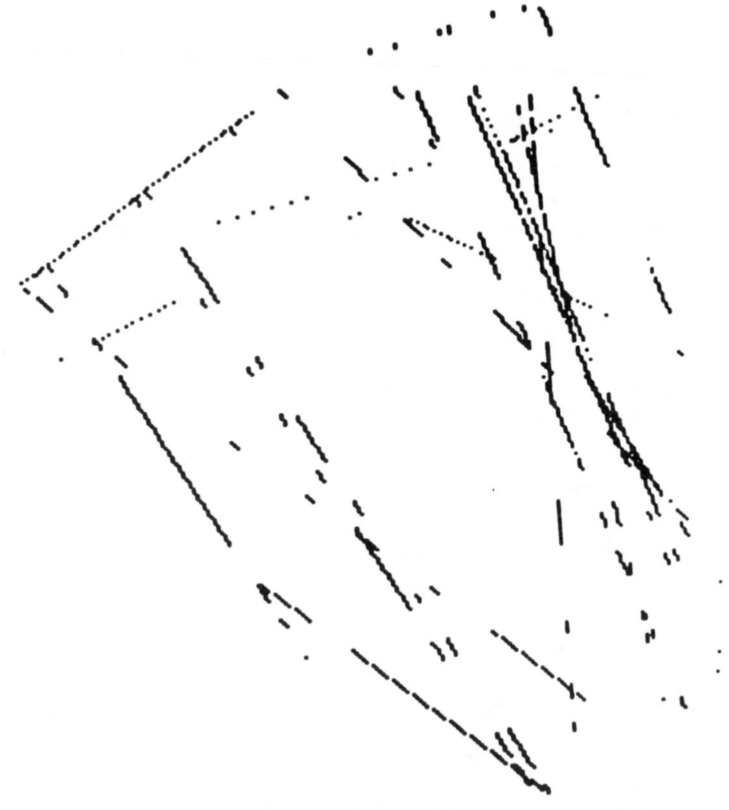

FIGURE 13 The heap-of-parts image after the Hough transform.

parameter space is quantized in the form of an array that is referred to as the accumulater array and is initialized to zeros. Then for each edge point in the image array, all accumulator array elements that are intersected by the line corresponding to the edge point are incremented by 1. Edge points that lie on straight lines in the image array intersect at the same location in the accumulater array and, hence, cause maximum accumulation. Therefore, if local maxima in the accumulater array are detected, the parameters of all straight lines in the image can be obtained. The values of the accumulater array elements give a measure of the number of points in each line.

The heap-of-parts image is shown in Figure 13 after the Hough transform.

High Level Processing

The objective of high level processing is generally the recognition of objects in the image. This is done through analysis, classification and identification of features extracted from the image during earlier low and intermediate level processing. Hence, high level processing may be generally characterized by analysis of feature lists that aim at the identification of objects in the image. The high level processing stage is basically the stage of image understanding.

High level processing and high level algorithms are, in general, less well understood and more difficult to characterize than low level and intermediate level processing and algorithms. This is basically attributed to:

1. High level algorithms generally involve symbolic processing of feature lists and/or employ techniques from several areas such as calculus, graph theory, differential geometry, category theory, logic and artificial intelligence. These techniques posses very diverse characteristics.
2. There is no uniform structure for the representation of the image in high level processing. In fact, high level algorithms vary considerably in this aspect. This is contrary to low level algorithms, for example, in which the representation of the image data is very uniform (two dimensional arrays of pixels).

We have, somewhat arbitrarily, chosen the following algorithms for our benchmark.

Graph Theoretic Algorithms

Graph theoretic algorithms refer to a class of high level processing algorithms that classify and recognize objects in the image by examining features extracted from the image against pre-defined models of the features of possible objects that can appear in the image[7]. A pre-defined model of the features of an object is referred to as the *template* of that object. That is, graph theoretic algorithms are matching algorithms in the general sense.

The features extracted from the image are represented as a directed graph (V,E) which consists of a set of vertices V and a set of edges E. The vertices represent the features extracted from the image. The edges represent relationships among these features. The template of an object is also represented in the same form. This representation of features is referred to as a *relational structure*[7].

Hence, the process for matching an object to a template can be formalized as one of the graph isomorphism related problems, such as graph isomorphism, subgraph isomorphism and double subgraph isomorphism.

In graph isomorphism, given the image data graph (V_1, E_1) and the template graph (V_2,E_2), the objective is to find a one-to-one onto mapping (i.e. an isomorphism) f between V_1 and V_2 such that for $v_1, v_2 \in V_1, V_2$ $f(v_1) = v_2$ and for each edge E_1 connecting any pair of vertices $v_1, v_2 \in V_1$, there is an edge E_2 connecting $f(v_1)$ and $f(v_2)$. In subgraph isomorphism, the objective is to find isomorphisms between the template graph and subgraphs of the image data graph. In double subgraph isomorphism, the objective is to find *all* isomorphisms between subgraphs of the template and subgraphs of the image data graph.

Graph isomorphism related techniques as described above are pure matching techniques. That is, the match between the template and the object should be perfect otherwise no-match is obtained. This can limit the usefulness of these techniques in the case of noisy input and/or imprecise templates. It is possible, however, in the implementation of the technique, to relax the strict rules of isomorphism correspondence and obtain a computational version of the technique that can take into consideration noise and imprecise templates. This is generally done by introducing matching metrics that measure the goodness of the match and hence, provide a quantified measure of matching. There are several methods for implementing

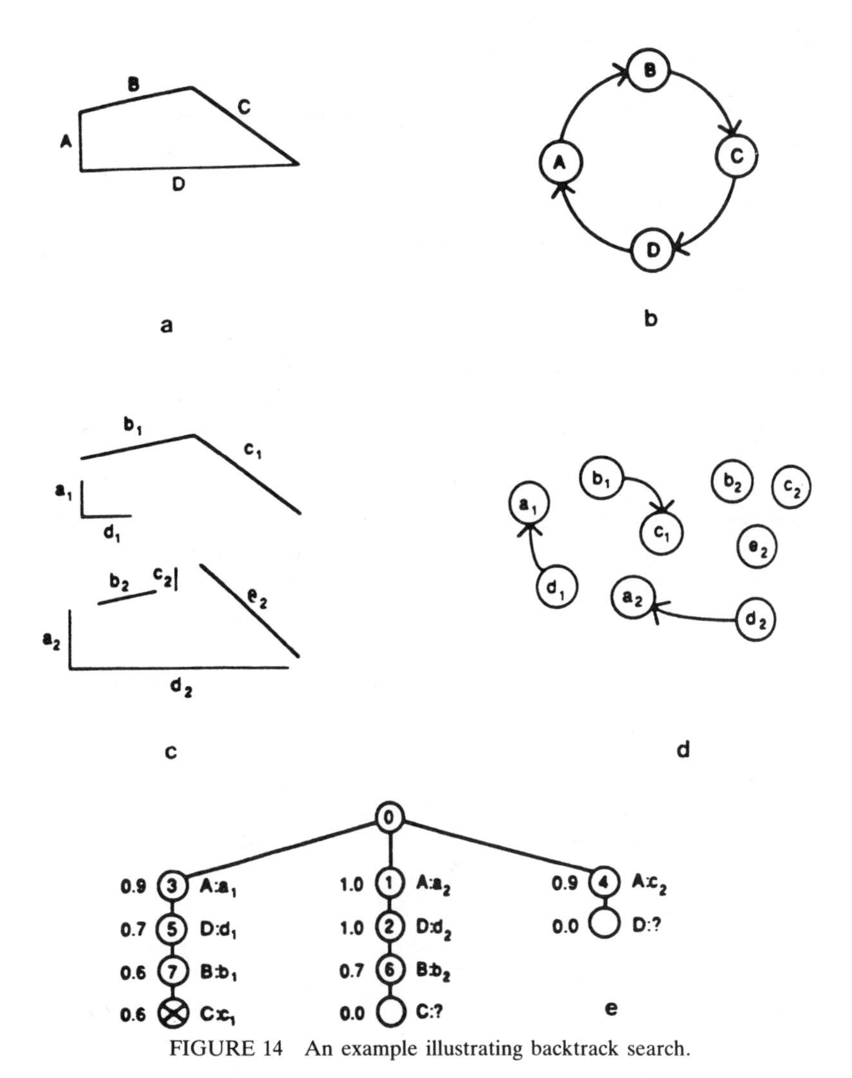

FIGURE 14 An example illustrating backtrack search.

graph isomorphism techniques with matching metrics. However, only the technique referred to as *backtrack search* is considered in this chapter.

The backtrack search technique refers to a set of possibly exhaustive search algorithms such as depth-first search, breadth-first search and best-first search. These algorithms differ in the order by which the search process proceeds. Therefore, only best-first search is described.

A generic version of the best-first backtrack search technique is illustrated using the simple example in Figure 14. The straight line segments extracted from the image and the relation 'next to' are used to construct the directed graph for the image data. This is shown in Figure 14(c), (d). The template of the object to be located and its pre-computed directed graph are shown in Figure 14(a), (b) respectively.

The backtrack search locates the object in the image in steps by matching a line segment in the template to a line segment in the image in each step until all line segments in the template are matched. A heuristic function is used to aid the search after each step by returning a value that indicates the goodness of the matches possible from that step. The heuristic function takes into consideration several factors such as how well the two straight lines match and how important the particular line segment in the template is in distinguishing the object in question from other objects that can appear in the image. This is referred to as the *saliency* of the line segment[8]. The search continues in the direction of the largest value returned by the heuristic function up to that point in the search.

The resultant search tree is shown in Figure 14(e). The letters to the right of each node in the tree indicate the two line segments matched at this node. The number to the left of each node is the value returned by the heuristic function. The integer inside each node indicates the order in which the nodes were expanded (decreasing heuristic function value). The search terminates when the object is located after matching C to c_1.

The Consistent Labeling Problem

The consistent labeling problem refers to the problem of assigning labels consistently to objects in the image. This problem is also known as relaxation labeling and constraint satisfaction[9]. Consistent labeling algorithms are generally employed after all objects in the image are identified.

The consistent labeling problem has four basic components: a set of objects identified in the image, a finite set of relations among the objects, a finite set of labels and a set of constraints that determine which labels may be assigned to which objects. Hence, the problem may be stated as: given the input as a relation structure (the objects and their relations), the labels and the constraints, assign labels to objects without violating the constraints.

The solution to the consistent labeling problem can be implemented in a number of ways[7]. It is also possible, in general, to employ backtrack search techniques to solve the consistent labeling problem[10,11].

ARCHITECTURES FOR ROBOT VISION

Research in special purpose architectures for Robot Vision has progressed considerably over the past decade. This has resulted in a large number of special purpose architectures for Robot Vision. With recent advances in computer hardware technology which is causing a steady decline in computer hardware cost, it is becoming feasible to build such special purpose architectures.

Special purpose architectures for Robot Vision can be divided

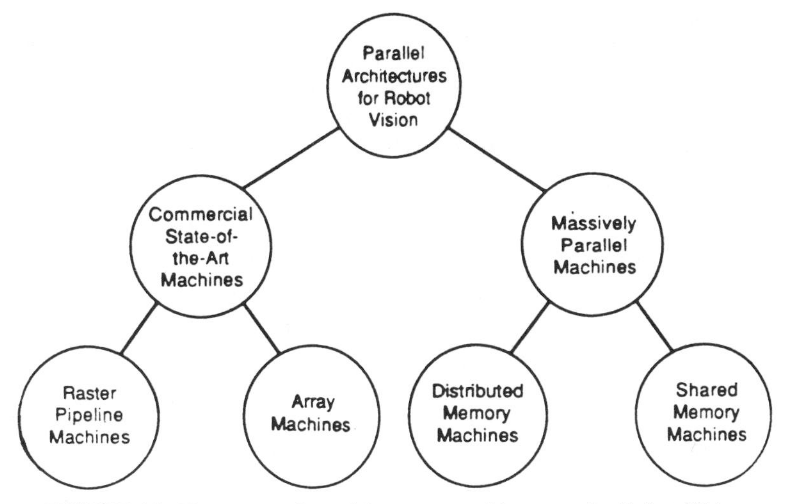

FIGURE 15 Taxonomy of special purpose architectures for Robot Vision.

into two major groups: *Commercial state-of-the-art* machines, and *Massively parallel* machines. The first group of architectures can be further subdivided into *Raster Pipeline* machines and *Array* machines. The second group of architectures may also be subdivided into *Distributed Memory* machines and *Shared Memory* machines. This taxonomy of special purpose architectures for Robot Vision is shown in Figure 15.

In the following sections, the basic features of each group of architectures are considered. The performance of each group on the set of benchmarks is also discussed.

Commercial State-of-the-Art Machines

The commercial state-of-the-art machines group of architectures refers to special purpose architectures that are currently available commercially. These machines are generally characterized by:

1. Low cost: the use of state of the art technology makes it possible to build these machines at low cost.
2. Most commercial state-of-the-art machines employ fixed point arithmetic. However, with advances in computer hardware technology, floating point arithmetic is starting to appear.
3. Machines that belong to this group of architectures function basically as attached processors to a host computer. The host computer, usually a conventional one, supervises the operation of the attached processor and performs functions such as loading and unloading of data and programs. The host computer also performs most I/O operations.
4. These machines are intended for Robot Vision applications which are generally simple in nature such as simple inspection of factory samples and aiding CAD graphics computations. There is also a reasonable amount of software available that implement these applications.

Raster Pipeline Machines

The basic architecture of Raster Pipeline machines is depicted in Figure 16. It employs a number of processing elements that are cascaded in series to form a pipeline of processing stages. The processing stages are controlled by the controller which issues instructions to the stages through a stage instruction bus during a setup phase and then streams the image data into the pipeline during the processing phase. The image buffer is used to hold the image data to be streamed into the pipeline as well as receive the output data from the pipeline. The host computer supervises the overall operation of the pipeline, the controller and the image buffer. Hence, the processing stages, the image buffer and the controller all function as an attached processor to the host computer [12].

Images can enter the pipeline in two ways: (1) the host sends the image to the image buffer where the controller streams it through the pipeline, back into the buffer and finally back to the host; and (2) the host sends and receives and image to and from the pipeline directly.

Images enter the pipeline as a steam of pixels in a sequential line-scanned raster format and move through the pipeline at a constant rate. Shift registers within each stage store two contiguous scan lines of the image. Furthermore, window registers, also within each stage, hold the pixels that form a 3 × 3 neighborhood. Each stage performs a pre-programmed function on this neighborhood. This function is referred to as the neighborhood function. The neighborhood function involves the transformation of the pixel in the center of the neighborhood based on that pixel value plus the values of its eight neighbors. The line and window registers are depicted in Figure 17.

At each discrete time step, a new pixel is clocked into each stage.

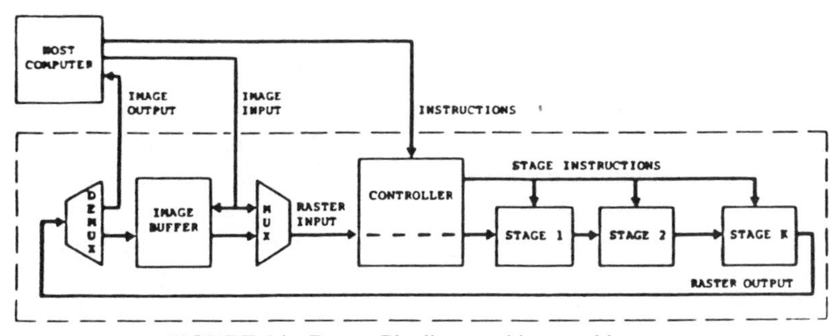

FIGURE 16 Raster Pipeline machines architecture.

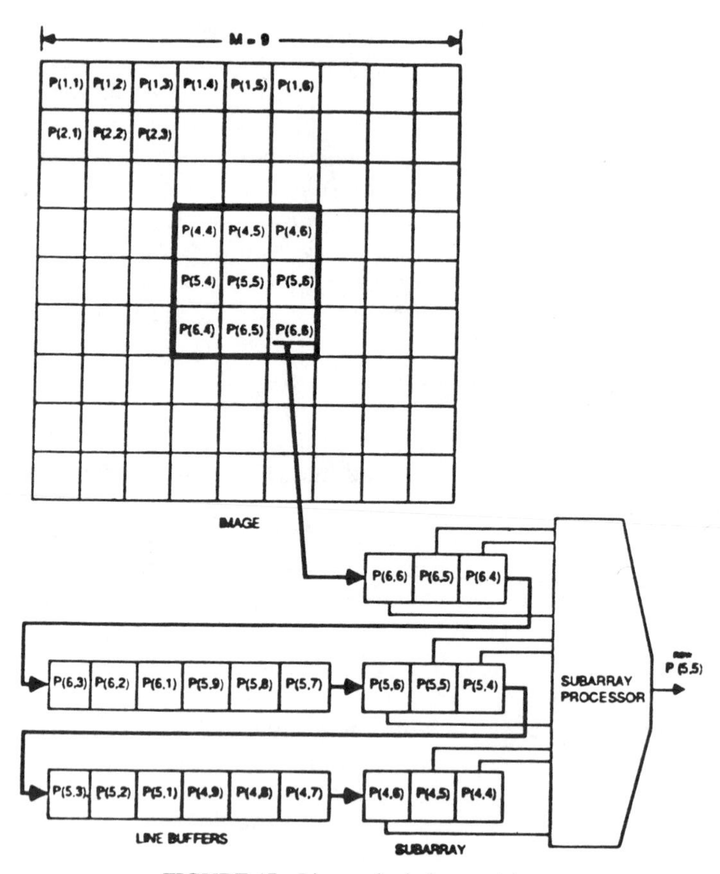

FIGURE 17 Line and window registers.

Simultaneously, the contents of all shift registers are shifted one element. Each stage performs the neighborhood function on the neighborhood it contains obtaining a new value for the center pixel. At the next time step, a new pixel is shifted into each stage and the transformed center pixel is shifted into the next stage. Hence, at each time step, each stage holds a new neighborhood in its window registers. The neighborhoods in three successive stages of the pipeline are shown in Figure 18. The output of the last stage forms the output stream of the pipeline and is fed back to the image buffer or directly to the host.

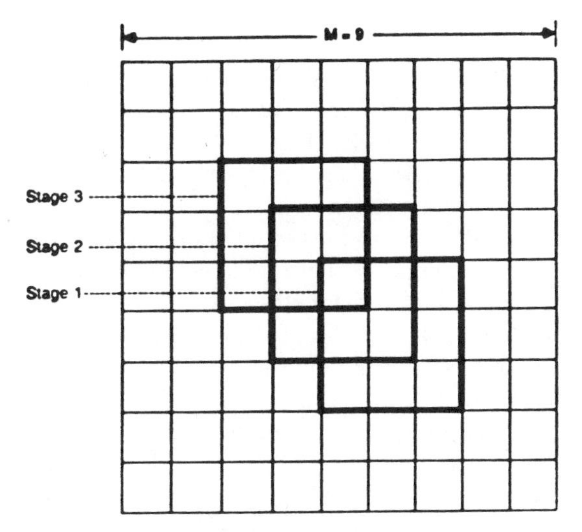

FIGURE 18 Neighborhoods in three successive stages of the pipeline.

Raster Pipeline processors offer a number of advantages:

1. Simple interconnections between the stages. This simplifies design and enhances reliability.
2. The input is in raster scan format which matches the output of many sensors. Hence, the image may be fed directly without reformatting for real-time processing.
3. At steady state, the output is obtained from the pipeline at a constant rate equal to that of the input to the pipeline.

However, there are several disadvantages for Raster Pipeline machines:

1. It is difficult to perform branching. The pipeline has to be flushed first.
2. It is difficult to process multiple images at the same time.
3. It is not possible to handle $1 \times m$ and $m \times 1$ neighborhoods that occur in FFT's and separable kernels.
4. They have a restricted set of instructions.

The cytocomputer is a typical example of Raster Pipeline machines[12,13].

Performance on Benchmarks

Low Level Processing. Assuming 16-bit fixed point arithmetic with 100 nsec multiply/accumulate time, Table 2 shows the times needed for convolving the image with an $m \times m$ FIR kernel. Each stage needs $m - 1$ line registers and m^2 window registers in order to be able to perform the convolution.

In order to perform the non-maximal suppression algorithm, the images containing the edge strength and the edge direction have to be streamed into the pipeline together, with a resolution of 8 bits per pixel. The non-maximal suppression algorithm involves less computations than a 3×3 convolution algorithm. Hence, the values for the 3×3 convolution algorithm in Table 2 are an upper bound for non-maximal suppression execution time provided additional hardware is available in each stage to implement the decision making required in this algorithm.

Intermediate Level Processing. The image array must be streamed through each stage of the pipeline K times, where K is proportional to the length of the longest sequence of connected pixels having strengths in the range $(T, \frac{T}{2}]$ before a strength T pixel occurs. That is, $K \propto \frac{1}{m}$, where m^2 is the number of window registers in the stage.

It is virtually impossible to implement the Hough transform algorithm on Raster Pipeline machines.

High Level Processing. The operations of high level processing algorithms are all virtually impossible to implement on a Raster Pipeline machine. Implementing search techniques, involved in high level algorithms, requires the capability of dealing with dynamic data structures. The design and instruction set of these architectures can not provide this capability.

Array Machines

The general block diagram of the Array machines architecture is depicted in Figure 19. It consists of an array processor (AP) which has its own local memory. The AP is generally a high speed processor that is optimized for vector and array operations. The AP communicates to the host computer through a high speed DMA channel. The image array is sent to the AP memory by the host over

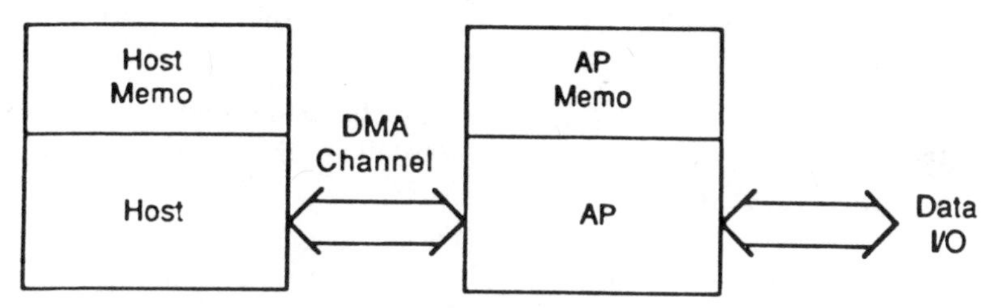

FIGURE 19 The architecture of Array machines.

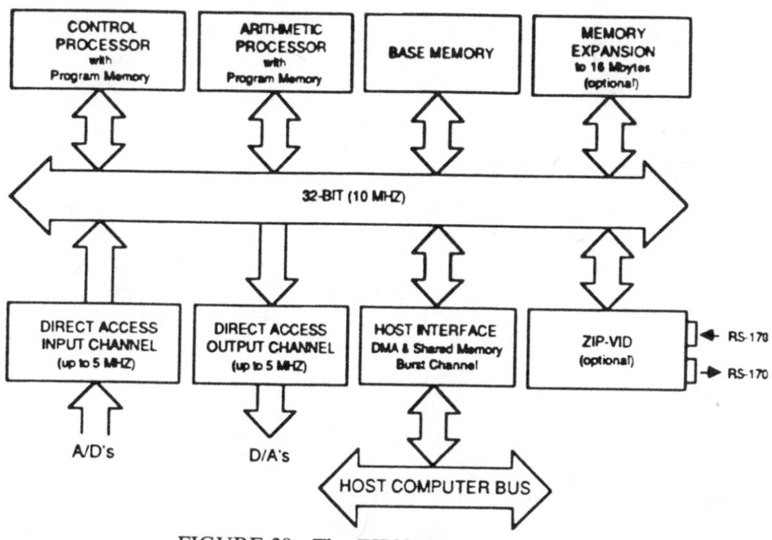

FIGURE 20 The ZIP3216 array processor.

the DMA channel. The AP then performs the required computation on the image array at high speed. The result is then sent back to the host over the DMA channel. Therefore, the use of the term *array* in this context reflects the nature of the data being processed.

Array machines offer several advantages:

1. Array machines offer more flexible programming than Raster Pipeline machines.
2. Array machines can easily handle large, symmetric and separable kernels.

3. The Array machine can operate in parallel with the host computer.

A typical example of the Array machines architecture is the ZIP3216 array processor[14]. The block diagram of the ZIP3216 is shown in Figure 20.

Performance on Benchmarks

Low Level Processing. A typical array processor, the ZIP3216, employs 16 bit fixed point arithmetic with 100 nsec multiply/accumulate time. Table 2 again shows the time for convolution for different kernel sizes.

	M = 256	M = 512
m	time (sec.)	time (sec.)
3	0.06	0.24
5	0.16	0.66
7	0.32	1.28
9	0.53	2.12
11	0.79	3.17

Table 2 Convolution times on Raster Pipeline machines

Intermediate Level Processing. The architecture of Array machines allows the processor to ignore non-edge pixels that are not needed for processing. In our experiments, the probability of a pixel being on an edge was typically in the range of 0.1−0.2. That is, less than 20% of the image has actually to be processed (20K to 50K bytes in a 256K byte image). Hence, for Array processors, the times for intermediate level processing is less than those for low level processing. This contrasts favorably with Raster Pipeline machines.

High Level Processing. Current Array machines do not provide the desired functionality for dealing with dynamic data structures needed for high level processing. In principal they could; however, this type of processing is intended to be done in the host and can be done concurrently with other processing.

Massively Parallel Machines

The Massively Parallel machines group of architectures refers to parallel architectures that employ a very large number of processors to achieve massive parallelism. The processors are interconnected in some manner to allow the sharing of data. Massively Parallel machines are generally characterized by:

1. High cost: due to the large number of processors and their interconnections, the cost of these machines is generally very high.
2. Massively Parallel machines are not widely available. They typically exist in research labs. In fact, many of these machines have yet to be implemented.
3. Major research must still be conducted in the areas of software and algorithm design for these machines.
4. I/O remains a bottleneck in these machines yet it has received very little attention.

The group of Massively Parallel machines is divided into two major subgroups: *Distributed Memory* machines and *Shared Memory* machines.

Distributed Memory Machines

In this group of architectures, each processor has its own local memory and the processors are interconnected to each other via communication links. The processors use these communication links to share data. The following are some examples of Distributed Memory machines:

Arrays

In this case, the processors are interconnected together to form a grid of processors. This is shown in Figure 21(a). Each processor is connected only to its four neighbors in the grid. The term *array*, in this context, is different from its use in the case of commercial state-

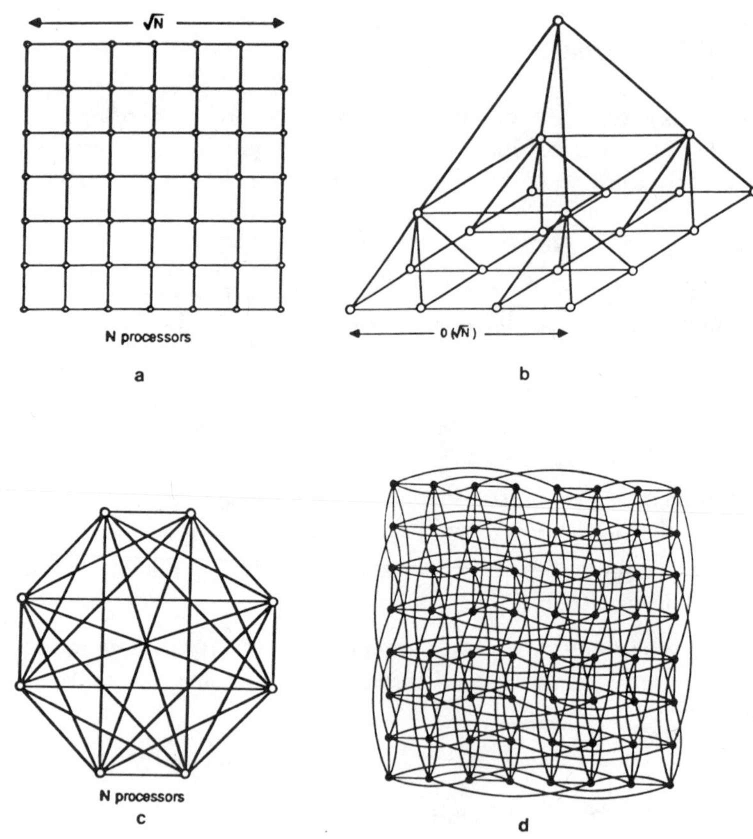

FIGURE 21 Distributed Memory architectures.

of-the-art machines and reflects the two dimensional spatial array of processors that form the grid. The advantage of this architecture is that the interconnection patterns are simple and are independent of the number of processors of the array. The disadvantage is that data movements across the grid beyond the four neighbors are time consuming ($O(\sqrt{N})$).

Examples of the array massively parallel architecture include the Illiac IV[15], which consists of 256 processors organized as four 8×8 grids of processors; and the Massively Parallel Processor MPP [16], which consists of a 128×128 grid of bit serial processors.

Pyramids

In this case, the processors are interconnected together to form a pyramid of processors. This is shown in Figure 21(b). Each processor is connected to four neighbors, a parent and four children processors. The advantage of this architecture is the simple interconnection pattern that is independent of the number of processors as well as the short distance between the processors; no two processors are more than $O(\log_2(N))$ steps apart. The disadvantage is congestion in the upper levels during system wide data transfers.

Examples of the pyramid massively parallel architecture include the PMPP[17] and the Pyramid Machine[26].

Completely Connected

In this case, every processor is connected to every other processor in the architecture. This is shown in Figure 21(c). Hence, each processor is exactly one step apart from any other processor. The advantage of this architecture is the high connectivity of the processors. The major disadvantage is the number of interconnections that grows quadratically with the number of the processors.

Hypercubes

In this case, the processors are connected to each other in such a way to form a hypercube array. This is shown in Figure 21(d) for 64 processors. The hypercube architecture offers several advantages:

1. The hypercube architecture is recursive. This makes the architecture recursively partitionable into smaller hypercubes.
2. It is possible to map the hypercube interconnection pattern into other important interconnection patterns such as grids and pyramids.
3. The hypercube architecture provides a good compromise between the decreasing proximity of processors and the number of interprocessor connections.

Examples of the hypercube architecture include the Intel iPSC, the NCUBE and the Ametek hypercubes. Table 3 summarizes and compares their basic features.

	Intel iPSC	Ametek 14/n	NCUBE/ten
No. nodes (max)	128	256	1024
Memory per node	512KB	1MB	128KB
CPU	286/287	286/287	Custom
	16 bit	16 bit	32bit
MIPS/node	0.30	0.30	2.00
MFLOPS/node	0.07	0.07	0.50
Node-to-Node	150KB/sec	N/A	1MB/sec
I/O Bandwidth	1MB/sec	N/A	90MB/sec

Table 3 Commercially available hypercubes

Shared Memory Machines

In this group of Massively Parallel architectures, each processor contains a limited size local memory and the processors are connected to a set of globally shared memory modules by an interconnection network. This is depicted in Figure 22. Hence, all processors share one common memory.

The interconnection network provides the means by which any processor may access any memory module as well as the means to resolve any conflicts that may occur on memory modules. There are numerous types of interconnection networks ranging from single shared buses to full crossbar switches[24].

The Shared Memory architecture has the advantage of providing a large shared memory for the processors. This is convenient for conventional block structured languages that imply a shared memory architecture such as Ada. However, as the number of processors increase, the interconnection network dominates the system cost and can degrade its performance.

Examples of interconnection network based machines include: PASM[18], which consists of 1024 processors and a reconfigurable multistage interconnection network; and the IBM RP3[19] which uses 512 processors.

The performance of Massively Parallel machines on the benchmark algorithms is presented below for hypercube architectures. The use of the hypercube architecture as a representative for Mass-

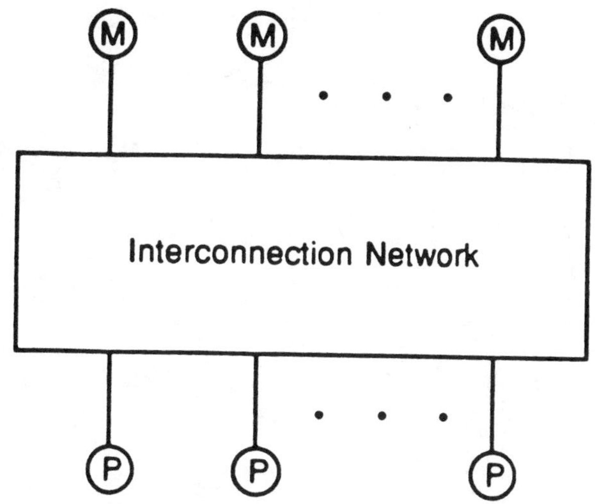

FIGURE 22 Shared Memory architectures.

ively Parallel architectures is motivated by the several advantages this architecture offers and the fact that a number of hypercube machines are already commercially available. A model for hypercube architectures is first presented. This model is then used to evaluate the performance of hypercube architectures on the benchmark algorithms.

A Model for Hypercube Machines

An n-cube array can be constructed recursively from $N = 2^n$ node processors as follows:

1. Basis Step: Form a 1-cube from 2 processors connected by a single communication link. Label one node 0 and the other 1.
2. General Step: Construct an n-cube from two $(n - 1)$-cubes as follows. First prefix the node labels in one of the $(n - 1)$-cubes with an 0. Second, prefix the node labels in the other $(n - 1)$-cube with a 1. Finally connect the two $(n - 1)$-cubes with communication links between pairs of nodes that have labels differing only in their most significant bit. A 4-cube (16 nodes) is shown in Figure 23.

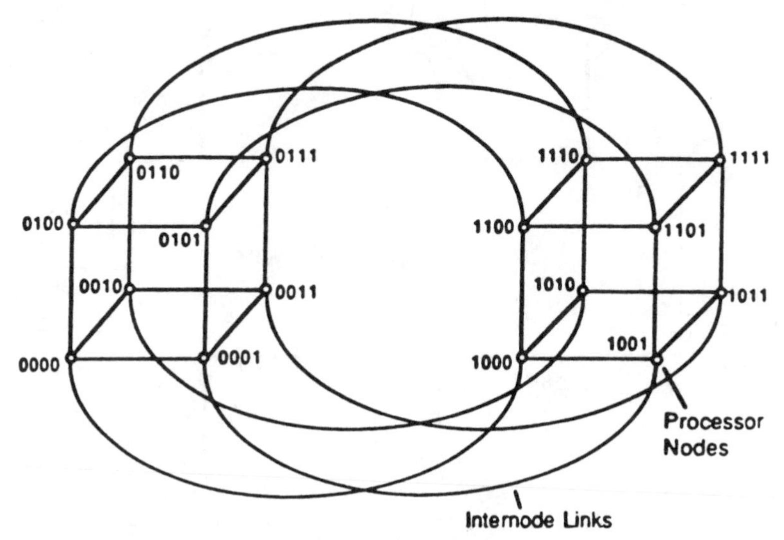

FIGURE 23 A 4-cube.

The cube array is connected via a set of I/O channels to a cube manager. This is shown in Figure 24. The cube manager is used for I/O (disks, tapes, cameras, sensory devices, etc.), as a peripheral controller and for program development. The cube array and the cube manager are referred to as the hypercube system.

A model for the node processor is shown in Figure 25. It consists of a CPU with a cache or large register file, main memory and $n + 1$ bidirectional DMA channels. The first n of the DMA channels are used to join the node processor to its nearest n neighbors in the cube array. The $(n + 1)^{st}$ DMA channel provides a link for communicating with the cube I/O system. It is assumed that caching within a node allows the DMA to proceed so that a fraction γ of the internode communication time can be overlapped with the node processing. The factor γ is referred to as the *degree of transparency*.

The time for an algorithm to run on a hypercube is given by:

$$T(N) = T_i + T_p(N) + (1 - \gamma)T_c(N) + T_o \qquad (1)$$

where N is the number of node processors in the cube array, T_p is the time of perform the processing, T_c is the internode communication time, T_i is the time to input the image data, T_o is the time to

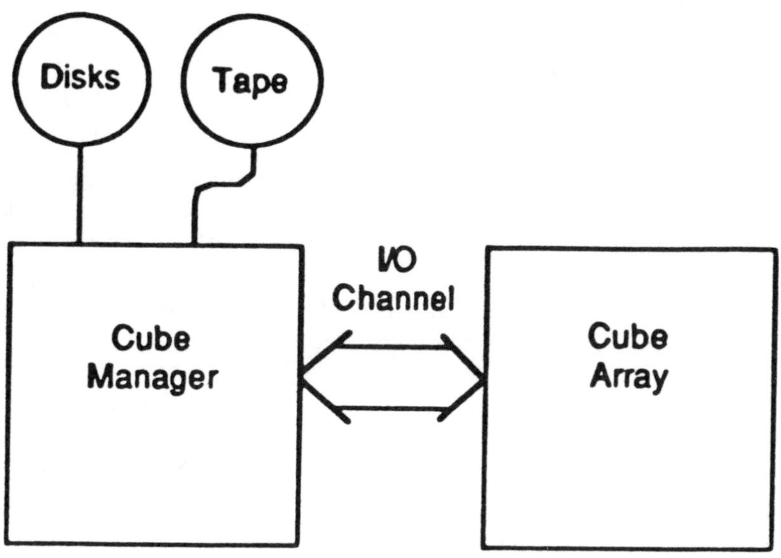

FIGURE 24 The hypercube system.

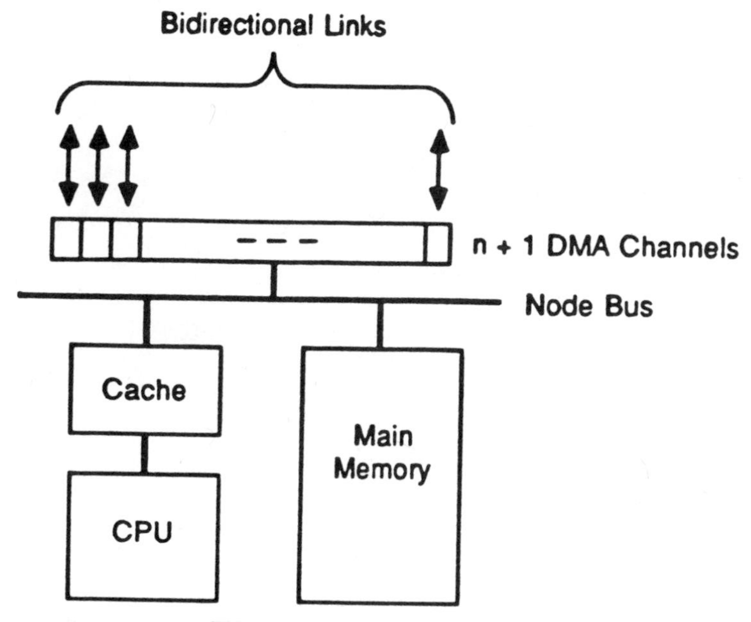

FIGURE 25 The node processor.

output the image data, and γ is the degree of transparency. Ignoring the I/O time, equation 1 becomes:

$$T(N) = T_p(N) + (1 - \gamma)T_c(N) \tag{2}$$

There are two principal contributors to intrinsic inefficiency of parallel algorithms on Massively Parallel architectures such as the hypercube. The first is the communication overhead which is incurred when $\gamma < 1$. The other is the dependencies within the algorithm that do not permit all N processors to be used all the time. This is reflected by the node efficiency, given by:

$$E_p(N) = \frac{T_p(1)}{NT_p(N)} \leq 1$$

The overall system efficiency is given by:

$$E(N) = \frac{T(1)}{NT(N)} \tag{3}$$

Using equations (2) and (3), and noting that $T_c(1) = 0$, then:

$$E(N) = \frac{E_p(N)}{1 + (1 - \gamma)\dfrac{T_c(N)}{T_p(N)}}$$

We define a *perfectly scalable* algorithm as one for which $E(N) = 1$, $N > 1$. This requires that $E_p(N) = 1$, i.e., processing is 100% efficient; and $(1 - \gamma)T_c(N) = 0$, i.e., the communication overhead is zero. Loosely speaking, a perfectly scalable algorithm can make use of large number of processors without diminishing returns. In general, it is desirable to design the parallel algorithm to be perfectly scalable[20].

Performance on Benchmarks

Low Level Processing

The image is partitioned into subimages of equal sizes, each subimage assigned to a node processor. A natural assignment for the

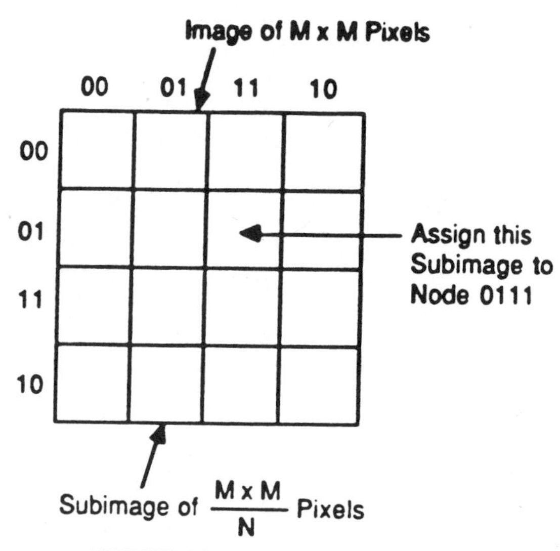

FIGURE 26 Partitioning the image.

hypercube is to partition the $M \times M$ image into a Gray code of $2^{n/2} \times 2^{n/2}$, (n assumed even), subimages similar to an n-dimensional Karnaugh map, and then to place each subimage with its like numbered processor. This is shown in Figure 26 for the hypercube shown in Figure 23. This method for partitioning the image and assigning the partitions to node processors guarantees that adjacent subimages are in adjacent processor nodes[20].

The algorithms in the benchmark for low level algorithms can be implemented as identical programs running in parallel, each in a processor node. That is, the hypercube is simulating an SIMD (Single Instruction stream, Multiple Data stream) machine. Hence, $E_p(N) \approx 1$ and the only potential contributor to inefficiency would be the communication overhead that results from the need to exchange data around the edges of the subimage to implement neighborhood operations. This is depicted in Figure 27. It shows a subimage in node A and the data that has to be moved from adjacent nodes. The number of pixels that has to be transferred is given by:

$$\frac{4M}{\sqrt{N}}\left\lfloor\frac{m}{2}\right\rfloor + (m - 1)^2$$

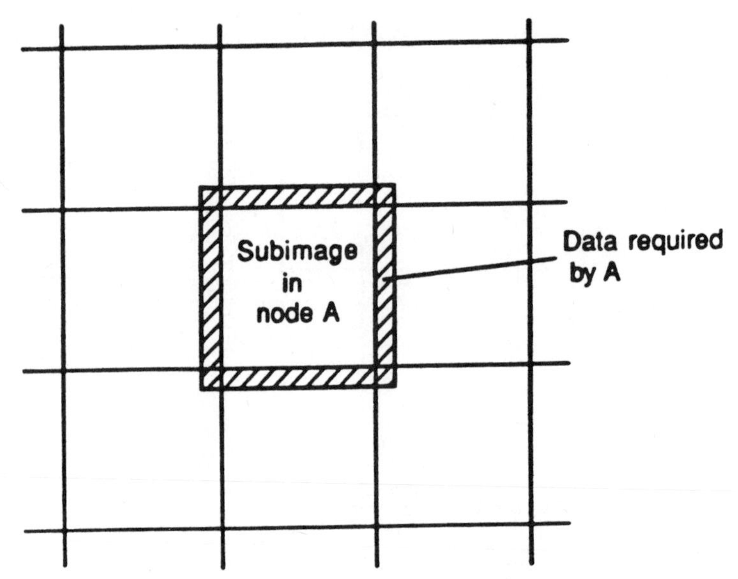

FIGURE 27 Data needed by a subimage in node A.

where $m \times m$ is the size of the kernel used and $M \times M$ is the size of the image. N is the number of node processors. The communication time necessary to move the pixels is proportional to their number. Hence:

$$T_c(N) = K_2\left[\frac{4M}{\sqrt{N}}\left\lfloor\frac{m}{2}\right\rfloor + (m-1)^2\right] \qquad (K_2 \textit{ is a constant}) \qquad (4)$$

The processing time at a node is proportional to the number of pixels in that node. All nodes have the same number of pixels. Hence:

$$T_p(N) = K_1 m^2 \frac{M^2}{N} \qquad (K_1 \textit{ is a constant})$$

To insure perfect scalability, the following is required:

$$E_p(N) = 1 \qquad (\textit{which is true})$$

and

$$(1 - \gamma)T_c(N) \ll T_p(N).$$

That is,

$$\frac{K_2}{K_1}(1 - \gamma)\left[\frac{2\sqrt{N}}{mM} + \frac{N}{M^2}\right] \ll 1.$$

The above equation suggests that the granularity of the subimages be fairly large if scalability of the algorithm is to hold.

Intermediate Level Processing

The image is mapped onto the nodes of the hypercube array in exactly the same manner as in low level processing. In fact, in many cases intermediate level processing directly follows low level processing and the image is already mapped in that manner. The two factors that affect the perfect scalability of the algorithm are node efficiency and communication overhead.

The subimages, which are of equal areas require, in general, unequal amounts of processing times in the case of intermediate level algorithms. This can be easily seen in the example of the edge following algorithm where each subimage contains a different number of edge pixels. This uneven workload results in some processing nodes finishing the algorithm before other processing nodes and, consequently, waiting idle for them. This in turn causes the node efficiency to drop and affects the perfect scalability of the algorithm. In [21], a Binomial stochastic model is assumed for the distribution of the edge pixels in the image. The model assumes that the probability of a pixel being on an edge is a constant p for all pixels in the image. The drop in efficiency is then quantified. The result in shown in Figure 28 for various values of p as a function of the number of pixels per subimage. As the number of processing nodes grows, the number of pixels per subimage decreases and the node efficiency drops. It might be necessary to re-map the image based on an equal load basis rather than on an equal size basis to improve the node efficiency. Further research is needed to evaluate possible re-mapping strategies.

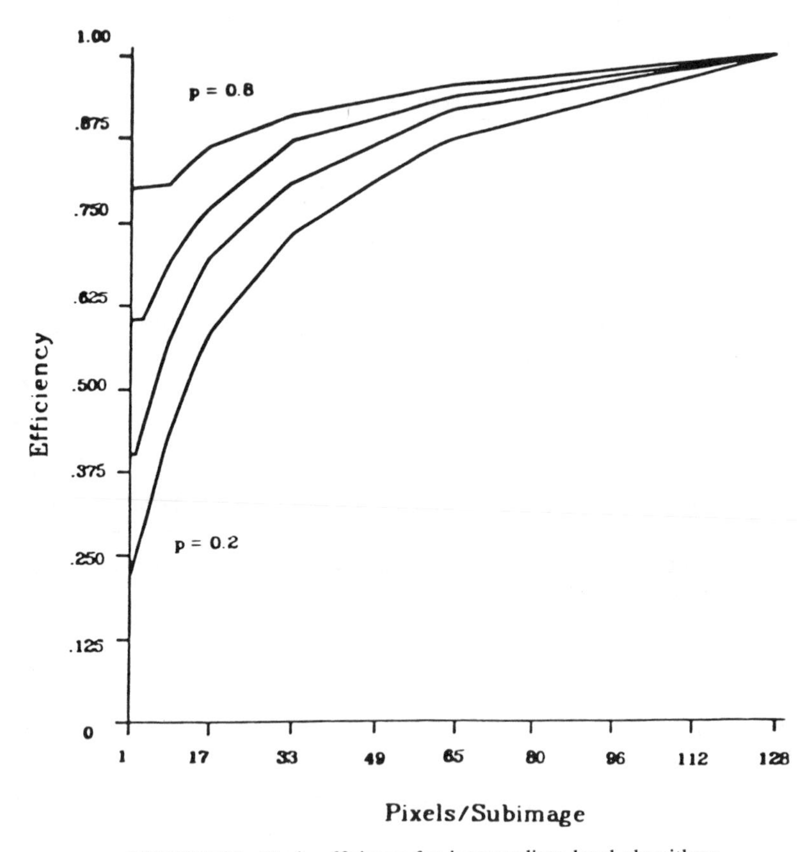

FIGURE 28 Node efficiency for intermediate level algorithms.

The communication overhead for intermediate level algorithms is, in general, difficult to estimate without knowledge of the distribution of the edge pixels in the image. However, intermediate level algorithms generally manipulate less pixels than low level algorithms and, hence, less pixels have to be communicated among the processing nodes. Therefore, equation 4 can form an upper bound on the communication overhead. However, further research still has to be conducted to accurately estimate the communication overhead for intermediate level algorithms.

High Level Processing

The implementation of some of the backtracking search algorithms for high level processing can be, in general divided into three phases: startup phase, computation phase and wind-down phase[22]. In the startup phase, the problem is assigned to a single node in the hypercube array. The problem is then expanded in that node and diffused to other nodes in the hypercube array. In the computation phase, the processors are busy performing the necessary computations. Finally, in the wind-down phase, the results are collected from the hypercube nodes and are combined to obtain the final result.

The node efficiency depends to a large extent on the ratio between the amount of time spent in the computation phase and the time spent in the other two phases. Research has yet to be conducted to quantify this ratio and, hence, the node efficiency.

The communication time overhead depends on the time spent in the startup and wind-down phases as well as whether the node processors communicate during the computation phase to maintain a balanced load of subproblems in the processing nodes. Again, further research is needed to determine the communication overhead.

CONCLUSIONS

In this chapter a study of special purpose architectures for Robot Vision was presented. The study was composed of two parts. In the first part, algorithms for Robot Vision were classified and the basic characteristics of each class were discussed. Benchmark algorithms for each class were proposed. Although the benchmark algorithms are only a sample of the possible algorithms used in Robot Vision, they do reflect the basic characteristics of each class of these algorithms.

In the second part of the study, special purpose architectures for Robot Vision were discussed. The basic features and characteristics of each group of architectures were presented. Furthermore, the performance of each group of architectures on the benchmark algorithm was presented.

While commercially available machines provide a reasonable solution to some low level processing, they do not provide enough computational power nor do they provide the desired functionality for intermediate and high level processing. It appears for future Robot Vision application that Massively Parallel machines will be needed to provide that power and functionality. However, considerable research has still to be conducted to determine how best to use these machines for higher level processing.

REFERENCES

1. M. Brady, *Computational Approaches to Image Understanding*, *Computing Surveys*, vol. **14**, no. 1, pp. 3–71, March 1982.
2. E.R. Davis, *Image Processing — its Milieu, its nature, and Constraints on the design of Special Architectures for its Implementation*, in *Computing Structures for Image Processing*, M.J. Duff, ed., New York: Academic Press, 1983.
3. P.E. Danielsson and S. Levialdi, *Computer Architectures for Pictorial Information Systems*, *Computer*, vol. **14**, no. 11, pp. 53–67, Nov. 1981.
4. A. Rosenfeld and A.C. Kak, *Digital Picture Processing*, New York: Academic Press, 1976.
5. R.C.Gonzalez and P. Wintz, *Digital Image Processing*, Reading: Addison-Wesley, 1977.
6. J.F. Canny, *Finding Edges and Lines in Images*, Master Thesis, Department of Electrical Engineering and Computer Science, MIT, June 1983.
7. D.H. Ballard and C.M. Brown, *Computer Vision*, New Jersey: Prentice-Hall, 1982.
8. J.L. Turney, T.N. Mudge and R.A. Volz, *Recognizing Partially Occluded Parts*, *IEEE Trans. Pattern Analysis and Machine Intelligence*, vol. PAMI-7, no. 4, pp. 410–421, July 1985.
9. B.A. Nadel, *The Consistent Labeling Problem, Part 1: Background and Problem Formulation*, Report DCS-TR-164, Computer Science Dept., Rutgers University, New Brunswick, N.J., 1985. Also appears as Report CRL-TR-13-85, Dept. Electrical Engineering and Computer Science, University of Michigan, Ann Arbor, MI 1985.
10. B.A. Nadel, *The Consistent Labeling Problem, Part 2: Subproblems, Enumerations and Constraint Satisfiability*, Report DCS-TR-165, Computer Science Dept., Rutgers University, New Brunswick, N.J., 1985. Also appears as Report CRL-TR-14-85, Dept. Electrical Engineering and Computer Science, University of Michigan, Ann Arbor, MI 1985.
11. B.A. Nadel, *The Consistent Labeling Problem, Part 3: The Generalized Backtracking Algorithm*, Report DCS-TR-166, Computer Science Dept., Rutgers

University, New Brunswick, N.J., 1985. Also appears as Report CRL-TR-12-85, Dept. Electrical Engineering and Computer Science, University of Michigan, Ann Arbor, MI 48109.

12. R.A. Rutenbar, T.N. Mudge and D.E. Atkins, *A Class of Cellular Architectures to Support Physical Design Automation, IEEE Trans. on CAD of IC's and Systems*, vol. CAD-3, no. 4, pp. 264–278, Oct 1984.
13. R.M. Lougheed and D.L. McCubbrey, *The Cytocomputer: A Practical Pipelined Image Processor, Proc. 7th Annual Symp. on Computer Architecture*, pp. 271–277, May 1980.
14. S. Krishna and R. Frisch, *Array Processor Tamed by Structural Innovations, Electronic Products Magazine*, Aug. 1984.
15. G.H. Barnes, et al., *The Illiac IV Computer, IEEE Trans. Computers*, vol. C-**17**, no. 8, pp. 746–757, Aug. 1968.
16. K.E. Batcher, *Architecture of a Massively Parallel Processor, Proc. 7th Annual Symp. on Computer Architecture*, pp. 168–174, May 1980.
17. D.H. Schaefer, *A Pyramid of MPP Processing Elements — Experiences and Plans, Proc. 18th Int'l Conf on System Sciences*, 1985.
18. H.J. Siegel, et al., *PASM: A Partitionable SIMD/MIMD System for Image Processing and Pattern Recognition, IEEE Trans. Computers*, vol. C-**30**, no. 12, pp. 934–947, Dec. 1981.
19. G.F. Pfister, et al., *The IBM Research Parallel Processor Prototype (RP3): Introduction and Architecture, Proc 1985 Int'l Conf on Parallel Processing*, pp. 764–771, Aug. 1985.
20. T.N. Mudge, *Vision Algorithms for Hypercube Machines, Proc of the IEEE Workshop on Computer Architecture for Pattern Analysis and Image Database Management*, pp. 225–230, Nov. 1985.
21. T.N. Mudge and T.S. Abdel-Rahman, *Efficiency of Feature Dependent Algorithms for the parallel Processing of Images, Proc. 1983 Int'l Conf on Parallel Processing*, pp. 369–373, Aug. 1983.
22. B.W. Wah, G.J. Li and C.F. Yu, *Multiprocessing of Combinatorial Search Problems, Computer*, vol. **18**, no. 6, pp. 93–108, June 1985.
23. A.V. Oppenheim and R.W. Schafer, *Digital Signal Processing*, Englewood Cliffs: Prentice-Hall, 1975.
24. T.Y. Feng, *A Survey of Interconnection Networks, Computer*, vol. **14**, no. 12, Dec. 1981.
25. R.O. Duda and P.E. Hart, *Use of the Hough Transform to Detect Lines and Curves in Pictures, Comm. ACM*, vol. **15**, no. 1, pp. 11–15, Jan. 1972.
26. S.L. Tanimoto, *A Pyramidal Approach to Parallel Processing, Proc. 10th Annual Int'l Symp. on Computer Architecture*, Stockholm, Sweden, pp. 372–378, June 1983.

Chapter 6

HIGH PERFORMANCE SPECIAL-PURPOSE COMPUTER ARCHITECTURES FOR ROBOTIC SENSING APPLICATIONS

Yuen-wah Eva Ma, Ramesh Krishnamurti, Bhagirath Narahari, Dennis G. Shea* and Kwang-shi Shu

Department of Computer and Information Science
Moore School of Electrical Engineering
University of Pennsylvania
Philadelphia PA 19104

The research objective of the REPLICA project is to investigate the design of high performance special-purpose computers for the support of robotic applications. In the design of special-purpose computers, the application characteristics must be carefully analyzed and thoroughly understood. Not only can such characteristics affect design decisions, they can also be used by the system to optimize dynamically its performance. The impact of application characteristics on design decisions is

---

This material is based upon work partly supported by National Science Foundation grant ECS 84-04741, National Science Foundation PYI award DCR84-51408, AT & T Information System research grant, National Science Foundation CER grant MCS82-19196, Army Research Office grant DAAG-29-84-K-0061. The United States Government is authorized to reproduce and distribute reprints for Governmental purposes notwith-standing any copyright notation hereon.

*The author is also affiliated with IBM Thomas J. Watson Research Center.

illustrated by the evaluation of interconnection networks for computers supporting robotic applications. The use of application characteristics in optimizing system performance is illustrated by the use of such information during the operation of a dynamically reconfigurable and partitionable computer for robotic applications to determine dynamically the optimal partition sizes.

INTRODUCTION

The increases in performance and sophistication of robotic systems are placing exceedingly high demands on the supporting computer systems. For example, the design of high performance computers for real-time machine vision remains a continuing challenge since current machine vision algorithms are computationally expensive. For robotic systems using multiple sensors, the demands on the performance of the supporting computers are much more stringent. These systems have to coordinate multiple sensors and process multiple streams of sensory data. The use of multiple sensors can make many tasks in robotic applications (such as object recognition) much easier by identifying different properties of the object, in addition to those obtainable through the visual sensor [1]. For example, some physical edges and holes are much easier to detect by a tactile sensor than visual sensor. The ambiguity due to shadows are easier to resolve by the use of a tactile sensor. However, such advantages cannot be realized unless the required system performance can be achieved by the supporting computers. The objective of this chapter is to examine the important issues in the design of high performance special-purpose computers for such support.

By focussing on a particular application, a special-purpose computer can be tailored to meet the performance requirements. For example, a system for robotic applications should be designed to handle large amounts of computations efficiently, while efficient handling of recursion might not be required. To design a special-purpose computer, the application characteristics must be carefully analyzed and thoroughly understood. These characteristics not only influence many design decisions (such as the choice of the interconnection network and the sizes of the memories in the memory hierarchy), they can also be exploited to optimize the system performance. Since a special-purpose computer is designed for a particular application, its performance should be evaluated by its effectiveness in supporting a specific set of tasks required by the application. Design decisions in such a computer may thus depart from a general-purpose computer.

Many conventional performance measures do not capture the application characteristics and hence are unsuitable and can even be misleading for evaluating special-purpose computers. Performance measures which capture both the characteristics of the application and the system are needed to evaluate such computers. Furthermore, the information on application characteristics not only can be used by the designers in making design decisions, they can also be used by the operating system to dynamically optimize the system performance during operation.

Active multi-sensory perception is a research project at the University of Pennsylvania [1,2,18], and REPLICA is a research project which investigates the design of high performance special-purpose computers for the support of robotic applications. In this chapter, we present our research results in the REPLICA project. The goal of our research lies in the integration of application characteristics in the design of the supporting computers. Our efforts are towards a design methodology which begins with capturing the application characteristics and then derives the desirable architectural features in the supporting computers. Preliminary results have been obtained but further research is needed to complete the design methodology.

In section 2, we discuss the impact of application characteristics in the design process. This impact is illustrated by the choice of interconnection networks for computers supporting robotic applications. The mesh, ring, and multi-stage interconnection networks (ICNs) are the three most popular ICNs for multi-processor systems. Conventional performance measures for these ICNs are first reviewed. Our proposed methodology in evaluating ICNs are then presented, which starts with classifying the problems in robotic applications and then deriving different performance measures for ICNs with respect to the classification. Conventional measures and our derived measures are then compared to pinpoint the deficiency of conventional measures in evaluating ICN for robotic applications.

In section 3, we examine the use of application characteristics in optimizing system performance. The design of a dynamically reconfigurable and partitionable system is used as an illustrative example, A dynamically reconfigurable and partitionable computer system provides the potential of achieving the system performance required by robotic applications using multiple sensors. One important management issue of such flexible systems is the choice of partition sizes in the system. Conventionally, this choice is either specified by the programmer or determined by the compiler. We show that by preanalyzing the task characteristics, the operating system can exploit this information and determine the partition sizes for various tasks dynamically based on the workload characteristics of the system during operation.

IMPACT OF APPLICATION CHARACTERISTICS ON ICN PERFORMANCE

Many multi-processor based parallel computers have been proposed and some implemented to support robotic applications. An important component of any multi-processor system is the ICN which supports communication among the processing elements (PEs). Since the ICN determines the communication delays in transmitting data among the communicating PEs, it directly affects the system performance in supporting parallel algorithms. Such communication delays are determined by two factors: (i) the inherent characteristics of the problems, and (ii) the inherent properties of the ICN. The inherent characteristics of a problem include its computation and communication structures, which determine the communication requirements on the system, and the inherent properties of an ICN include its topology and bandwidth, which determine the communication support of a system.

Conventional performance measures of an ICN include the band, width and the worst-case and average-case delay (defined respectively in terms of the worst-case inter-processor distance and average-case inter-processor distance among all the PEs in the system). Such measures only capture the inherent characteristics of an ICN but not the application, and therefore, they may not be accurate in evaluating ICN performance for special-purpose computers supporting robotic applications. For example, using either the worst-case delay or average-case delay, the multi-stage ICN has $O(\log(N))$ delay, the mesh has $O(\sqrt{N})$ delay, and the ring has $O(N)$ delay, where N is the number of PEs in the system, and accordingly, the mesh outperforms the ring and the multi-stage outperforms both. (The definitions of these ICNs are given in the next subsection.) However, for some particular problems in robotic applications (such as the linear scaling of an image) all three ICNs have the same delay, and for some others (such as the detection of horizontal lines) both the mesh and ring have shorter delay than the multistage. The key point is that for a special-purpose computer for robotic applications, performance measures which capture both the characteristics of the applications and the ICNs are needed to achieve an accurate evaluation of ICN performance.

Our methodology in evaluating ICN performance begins with classifying the problems in robotic applications according to their communication requirements, and then using suitable measures to evaluate ICN performance with respect to such a classification. In the following, we present the preliminary results in our methodology. Our current classification only applies to a subset of the problems in robotic applications, and it captures only the communication but not the computation characteristics of the problems. Our evaluation of ICN performance includes the mesh, ring, and multi-stage. For several classes of the problems in this classification, we are able to accurately evaluate these ICNs performance (in terms of interprocessor distance among communicating PEs to support the problem). Further research, however, is required to complete this evaluation and to extend the classification to capture the computation characteristics of the problems as well as to evaluate a wider spectrum of ICNs. In the following subsections, the term *delay* is defined to be the distance between communicating processors.

Basic Characteristics of the Mesh, Ring, and Multi-stage ICNs

The mesh, ring, and multi-stage are the three most popular classes of ICNs for robotic applications. Examples of those systems using

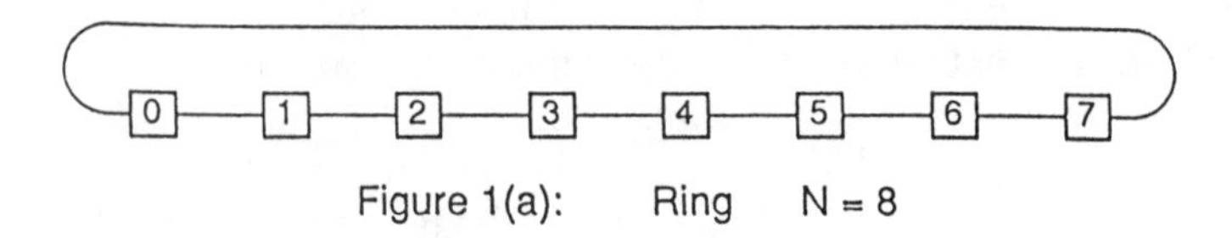

Figure 1(a): Ring N = 8

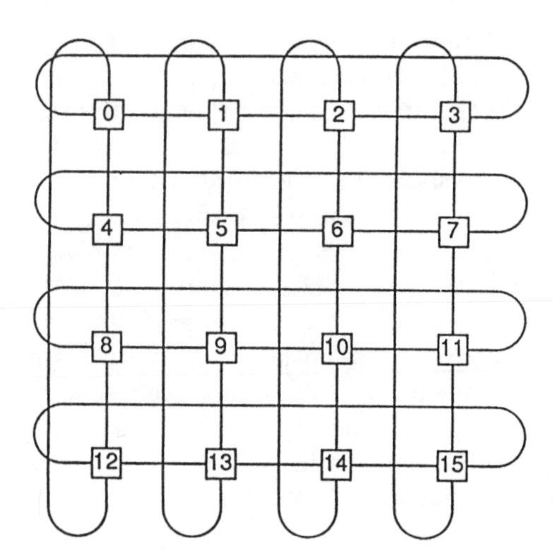

Figure 1 (b): Mesh for N = 16

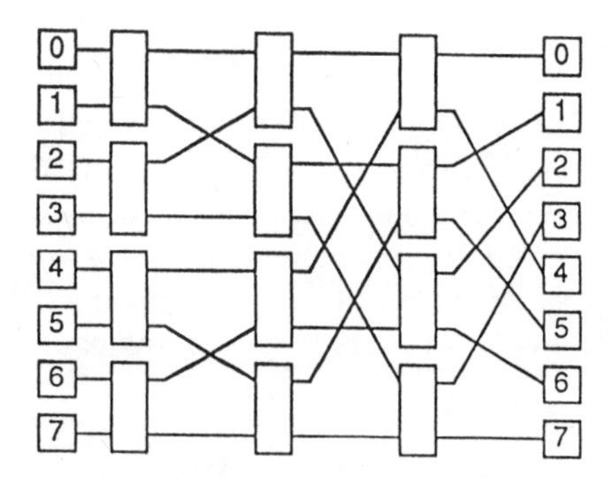

Figure 1(c): Multi- stage for N=8
(indirect binary n-cube)

the mesh include MPP [3] and GRID [17], those using the ring include ZMOB [10] and REPLICA [13], and those using the multi-stage include PASM [20] and PUMPS [5]. There are many variations in each of these classes of ICNs. In this chapter, we use the term mesh for the two dimensional (2-d) grid topology with wrap-around connections in the boundary PEs and the term ring for the bidirectional ring. To simplify our analysis, we ignore the blocking probability in the multi-stage, and hence its exact topology is not important in our analysis. The topologies of these three ICNs are given in *Figure 1*.

An important difference between the multi-stage and the mesh or ring is that the multi-stage has *homogeneous* communication delay (the same amount for every pair of PEs) while the mesh and ring have *non-homogeneous* communication delays. Since many problems in robotic applications have non-homogeneous communication requirements, the property of non-homogeneous communication delay in the mesh and ring can be exploited. Consequently, the mesh and ring can support many problems in robotic applications more efficiently than the multi-stage, despite their long worst-case and average-case communication delays.

Basic Characteristics of Problems in Robotic Applications

A common characteristic in many problems for robotic applications is that their input data are represented in the 2-d grid structures (2-d images). Furthermore, these problems require the application of a sequence of similar operations in each of the input data points. The problems in robotic applications are modeled by the following characteristics:

1. The size of the problem is much larger than the number of PEs in the system.
2. The input data is represented by a 2-d square grid structure. Multiple grids are also allowed.
3. The problem involves the execution of a sequence of similar operations on different subsets of input data points.
4. The problem is implemented by an SIMD algorithm (the synch-

ronization is implicit). This characteristic is assumed to simplify our presentation, while most of our results also apply to MIMD algorithms.

Classification of Problems in Robotic Applications

For any integer k, let $[k]$ denote the set of integers $\{0,\ldots,k-1\}$. A *dependency set* is defined to be a set of input data points which are all needed to compute some particular output point. Hence, a problem has the same number of dependency sets as the output points. Let M denote the number of input data points, and P denote the number of output data points. Let D_i denote the dependency set corresponding to output data point i, for $i \in [P]$. Based on their dependency sets, we classify problems into four classes: point dependency, fixed dependency, distributed dependency, and global dependency.

1. Point dependency: A problem has point dependency if $max_{i\in[P]} |D_i| = 1$. That is, each output data point can be computed from one distinct input data point. Linear scaling of pixel points of an image belongs to this class.
2. Fixed dependency: A problem has fixed dependency if $max_{i\in[P]} |D_i| = c$, where c is some constant and $c > 1$. Edge detection of an image belongs to this class.
3. Distributed dependency: A problem has distributed dependency if $max_{i\in[P]} |D_i| = f(M)$, where $f(M)$ is an increasing function of M and $f(M) < M$. An example in this class is the multiplication of two matrices (each of size $\sqrt{M} \times \sqrt{M}$). In this problem, each output data point in the product matrix is computed from $2\sqrt{M}$ input data points.
4. Global dependency: A problem has global dependency if $max_{i\in[P]} |D_i| = M$. That is, it has at least one output data point which requires all the input data points. Histogram computation of an image is such an example.

For problems with fixed dependency, we introduce another criteria called *locality* to further classify these problems. Let I denote the set of all input points. We have $|I| = M$. Since we assume 2-d input data, for any $X \in I$, the position of X in the 2-d representation can be denoted by (x_1, x_2), where $x_1, x_2 \in [\sqrt{M}]$. For any integer k, let $|k|$ denote the absolute value of k. Let $\delta: I \times I \rightarrow Z$ be a distance measure such that for any two input points X and Y in I, $\delta(X, Y) = |x_1 - x_2| + |y_1 - y_2|$. This distance measure δ is equivalent to the $l_p - norm$ (with $p = 1$) in functional analyses. For any integer k, let $|k|' = \min \{|k|, \sqrt{M} - |k|\}$. Let δ': $I \times I \rightarrow Z$ be the distance measure such that for any two input points X and Y in I, $\delta'(X,Y) = |x_1 - x_2|' + |y_1 - y_2|'$. This distance measure δ' measures the *modulo* distance in the 2-d input representation.

The locality of a dependency set is defined to be the maximum distance (defined by either δ or δ') among all pairs of input points in the set. Let $l_\delta(D_i)$ and $l_{\delta'}(D_i)$ denote the locality of D_i, where $i \in [P]$, as defined by δ and δ' respectively. We classify a problem to have *local dependency* with respect to δ if it has fixed locality and $max_{i \in [P]} l_\delta(D_i) = c$, for some positive constant c. The definition of a problem with local dependency with respect to δ' can be defined similarly. It should be noted that all the problems with local dependency with respect to δ are also classified as local dependency with respect to δ', but not the converse. Our following analyses of ICNs performance for problems with local dependency used δ' as the distance measure. However, similar conclusion can also be derived if δ is used instead. In the rest of this chapter, we use the term local dependency to mean the locality as defined by δ' without further specification.

Performance Evaluation of ICNs with respect to the Classification

It is obvious that for problems with point dependency, any ICN can support them with equal efficiency. For problems with global dependency, the worst-case communication delay in an ICN may dominate its efficiency. Intuitively, the mesh should outperform the ring, and the multi-stage should outperform both. However, in many of these problems, the large amount of computations and transfers can dominate over the worst-case delay. Due to their capabilities in overlapping communication and computation during data transfer, the performance of the mesh and ring compares favorably with the multi-stage for some of these problems. Such an example is given in [14].

For problems with local dependency, the ICN performance depends on its topology instead of the worst-case delay. Some of these problems can be implemented by a parallel program with unit delay (between communicating PEs) on one ICN but with much higher delays on another. Problems with local dependency have nonhomogeneous communication requirements, and thus, the ring and the mesh can exploit this property. However, due to the homogeneous communication delays in the multi-stage, it has $\Theta(\log N)$ delay for all these problems.†

Due to the similarity between the 2-d data representation and the mesh topology, for problems with local dependency, the input points can be mapped onto the mesh such that constant communication delays can be supported. Furthermore, if each data dependency set of the problem consists of an input data point and its four immediate neighbors as defined in the 2-d layout, the mesh can support it with unit delay, and the ring can support it with a delay of $\sqrt{N}$. (The delay $\sqrt{N}$ is optimal in the ring, that is, it is both an upper and lower bound in supporting problems with such dependency.)

Similarly, if the points in each dependency set are within a constant distance along one fixed dimension, the ring can support it in constant communication delay. Furthermore, if each dependency set of the problem consists of an input data point and its two immediate neighbors along one fixed dimension in the 2-d layout, the ring can support it with unit delay. If each dependency set of the problem consists of an input data and its four nearest neighbors along one dimension (that is, those input points within a distance of two along one dimension), the ring can support it with a delay of two and the mesh cannot support it with shorter delay than the ring. Hence, for problems with such dependency, the ring is at least as efficient as the mesh.

In general, the mesh has constant delay for all the problems with

---

†$O(g)$ is the set of f such that there exits a constant c where for all n, $f(n) < cg(n)$. $\Theta(g)$ is the set of f such that there exists a pair of constants c_1 and c_2 where for all n, $0 < c_1 g(n) < f(n) < c_2 g(n)$.

local dependency, the ring has $O(\sqrt{N})$ delay for all and constant delay for some, while the multi-stage has $\Theta(\log N)$ for all. A formal analyses for all the above-mentioned results are given in [14].

For problems with distributed dependency, the ICN performance may relate to the worst-case and average-case delays, and for problems with fixed dependency but not constant locality, the ICN performance may relate to its topology. However, further studies are required to analyze the ICNs performance more precisely.

Summary

Using either the worst-case delay or average-case delay, the multistage outperforms both the mesh and the ring. However, as we have illustrated in the above evaluation, these measures are not suitable for evaluating the ICN performance in supporting problems with either point or local dependency. The characteristics of the problems must be incorporated in achieving an accurate evaluation. Our current research includes the extension of the classification to capture the computation characteristics of problems and the evaluation of a wider spectrum of ICNs.

THE USE OF APPLICATION CHARACTERISTICS IN OPTIMIZING SYSTEM PERFORMANCE

Various multi-processor architectures have been developed for computer applications requiring high performance. These architectures include SIMD (single-instruction-stream, multiple-data-streams), MIMD (multiple-instruction-streams, multiple-data-treams), MSIMD (multiple SIMDs), and MSIMD/MIMD (multiple SIMD/MIMDs) systems. Robotic applications require a large amount of computations for both low-level sensory data processing and high-level object recognition. Algorithms for low-level sensory data processing are characterized by their simple deterministic processing applied to all the input data and such algorithms are most suitable to be performed by SIMD computers. Algorithms for high-level object recognition are less deterministic and apply to a smaller portion of the data and such algorithms are most suitable to be performed by MIMD computers.

An MSIMD/MIMD computer can dynamically partition itself into multiple SIMD and MIMD subsystems (called *partitions*). Furthermore, the system can reconfigure in the number of partitions and the configurations of the partitions (that is, either SIMD or MIMD). Such a system is particularly suitable for robotic applications using multiple sensors. For example, to perform object recognition in such applications, the type and amount of different sensory data needed to be processed varies from object to object, and thus, the processing requirements needed to handle these sensory data vary accordingly. The reconfigurability and partitionability of an MSIMD/MIMD system allows it to allocate computing resources to different partitions according to processing requirements, thus achieving efficient sharing of computing resources among the various sensors, and the ability of forming both SIMD and MIMD configuration makes it efficient in executing algorithms for both low-level sensory data processing and high-level object recognition. An MSIMD/MIMD system is reconfigurable and partitionable. In the rest of this chapter, we use the term partitionable computer interchangeably with the term MSIMD/MIMD system.

In this section, we examine the use of application characteristics in optimizing the performance of a special-purpose partitionable computer for robotic applications. We show that information on the application characteristics can be used by the operating system of a partitionable computer to determine the partition sizes for different tasks dynamically. This capability can improve the throughput of the system. In the following subsections, we first review and compare the properties of SIMD, MIMD, MSIMD, and MSIMD/MIMD systems. The relationship between partition sizes and the throughput of a partitionable system is then discussed. The impact of task and workload characteristics on optimal partition sizes are then illustrated. We then outline our results in designing an operating system that uses the information on application characteristics to optimize system performance.

Comparisons of Multi-processor Architectures

An SIMD or an MIMD system consists of a number of PEs and a control unit. In an SIMD system, the control unit broadcasts intructions to the PEs, which execute the instructions in lock-step. The PEs in an SIMD system are implicitly synchronized. This fully synchronized property simplifies the control of the PEs' activities. However, since the system can only execute one instruction stream at any time, the main problem in an SIMD system is that if the program does not have enough data streams, some PEs in the system have to be idle, and this may lead to a severe degradation in system performance. This problem of matching the number of data streams in a program to the number of PEs in the system is called *vector fitting*. In an MIMD system, each PE can execute different codes, and the activities of the PEs are coordinated by the control unit. In such a system, a program can be decomposed into independent subtasks which can be executed in parallel by the different PEs. This property make the architecture more flexible than an SIMD system, and hence, can be used to execute a wider class of programs. However, explicit synchronization is required among co-operating PEs. Therefore, the co-ordination of the activities of the PEs in an MIMD system is much more complicated than in an SIMD system.

An MSIMD system consists of a number of PEs, a number of control units, and a system control unit [16]. The set of PEs is shared by the control units. This system can dynamically partition itself into multiple numbers of independent subsystems, each consisting of a control unit and a number of PEs. These subsystems are called partitions. The number of partitions and the sizes of the partitions can vary, and are controlled by the operating system (executing in the system control unit). Such systems can execute several tasks in parallel, each on an SIMD subsystem. The partitionability of an MSIMD alleviates the problem of vector fitting, since the system can allocate a suitable number of PEs to a task according to its requirement. MAP is an example of MSIMD computers [16].

An MSIMD/MIMD system is similar to an MSIMD system except

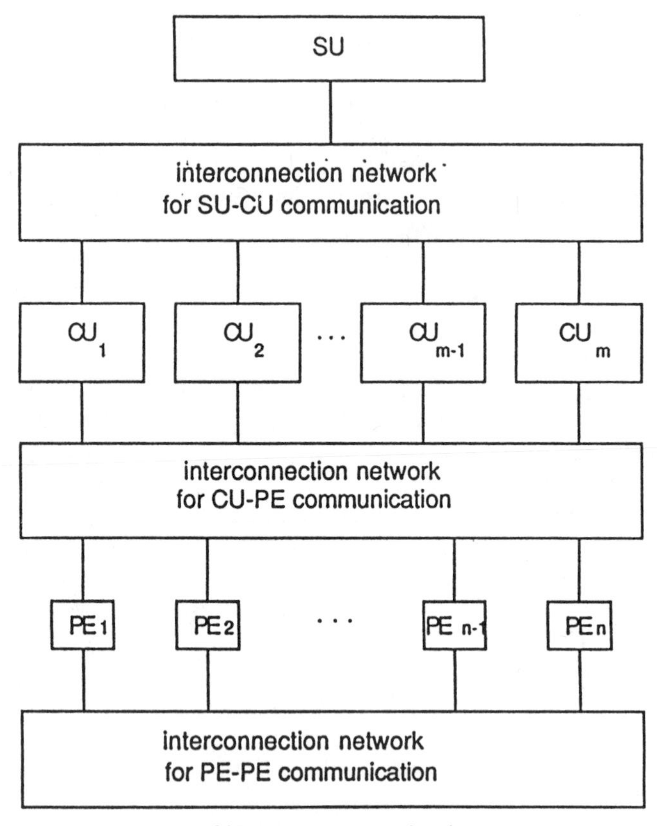

Figure 2: Block diagram of a partitionable computer system

that each partition can be an SIMD or an MIMD subsystem [20]. This option makes the system more flexible, and also combines the advantages of both SIMD and MIMD systems. These systems can process multiple jobs in parallel, each of which can be executed by an SIMD or an MIMD subsystem. *Figure 2* gives a block diagram of such systems. Some example systems of MSIMD/MIMD include PASM [20,9], TRAC [7,19], PM4 [6], and MOPAC [11].

A major limitation in both SIMD and MIMD systems is that, due to communication overhead, as we increase the number of PEs in the system the performance improvement of the system levels off [4,12,21]. Furthermore, in an SIMD system, since there is always a limit to the degree of parallelism in any particular task, beyond a certain number of PEs in the system further increase in this number will not lead to any additional performance improvement. This problem also occurs in an MIMD system, due to the increasing amount of control overhead in coordinating the activities of the PEs. On the other hand, since an MSIMD or MSIMD/MIMD system can dynamically partition itself into multiple independent subsystems, each of which can be of smaller size, it is possible to remove the above limitation of SIMD and MIMD systems. The ability in removing this limitation depends on the effectiveness in partitioning the system.

The Partitioning Issue

The effectiveness in partitioning a system depends on the following issues:

1. When to form partitions — Should the operating system initiate a new task as soon as there are enough PEs to form the partition, or should the system wait until all the tasks currently being executed are finished?
2. How to determine the partition sizes — Should the system rely on programmers to specify the sizes of the partitions explicitly in their programs, on the compiler to determine these sizes by analyzing the characteristics of the program, or on the operating system to determine these sizes dynamically based on the system characteristics and the workload at run-time?
3. How to form the partitions — Which PEs and control unit should be included in a partition?

The above three issues are interrelated, and they should all be taken into account in order to effectively partition a system.

Our research effort in designing partitionable systems focusses on the use of application characteristics to assist the system in choosing the optimal partitions. Our objective is to design an operating system that determines the partition sizes dynamically, based on the dynamic characteristics of the workload and the characteristics of the system at run time. The premise is that by determining the partition sizes dynamically, the performance of the system can be increased over that of a partitionable system in which partition sizes are chosen statically.

The impact of partition sizes on the throughput of a system has been discussed in [21,8]. Siegel et. al. showed that to execute a given task on M different data sets on a partitionable system with N PEs, the throughput (number of jobs executed per unit time) varies depending on whether all N PEs are used to execute the task on one data set at a time (that is, sequentially among the data sets), or only N/M PEs (assuming N is divisible by M) are used on each data set and executing all the data sets simultaneously [21]. Similar conclusion is drawn from a simulation study of the histogramming algorithm [8].

In the following subsections, the relationship among workload characteristics, task characteristics, and optimal partition sizes are presented, and the impact of these factors on system performance is analyzed.

Impact of Application Characteristics on Optimal Partition Sizes

The optimal partition size for a task is defined to be that partition size (the number of PEs in the partition) which can be used to execute the task with the shortest completion time. With respect to a specific workload, which consists of a number of tasks, *the optimal partition sizes for the tasks with respect to the workload* is defined to be those partition sizes which can be used to execute these tasks with the shortest overall completion time for the tasks. The optimal partition size for a task depends on its characteristics as defined by its computations and communications structures, and the optimal partition sizes for the workload depend on the characteristics of all the tasks in the workload. Hence, the optimal partition size for a task may not be the optimal partition size for the task with respect to the workload. This concept is illustrated by the following simple examples of summing and sorting.

In the following examples, we assume that we have a partitionable system in which the PEs are interconnected by a set of rings. Each partition is formed by a control unit and a number of PEs, where the PEs communicate using a dedicated ring (from the set of rings in the system), and the control unit communicates with the PEs using a dedicated link (which is part of a capability-enhanced cross-bar). This architectural model is based on the one proposed in [13]. To simplify the presentation of the analyses, we assume that the PEs in a partition are consecutive, and the time to send a data item between adjacent PEs equals the time to perform an addition or a comparison operation. While the data derived in the following examples depends on these assumptions, similar conclusions can be drawn with other models of partitionable architectures.

Let M denote the size of the task, K denote the size of the partition, and T denote the completion time of the task(s). We assume that the system has 512 PEs. To simplify the analysis, we assume that M is divisible by K (where $K \leq M$).

Impact of task characteristics on optimal partition size

Example 1: Summing M numbers

Assume that the task uses a recursive doubling algorithm to sum the M numbers, which takes $M/K - 1 + \log K$ additions and $K - 1$ communication operations. Hence, we have

$$T = \frac{M}{K} + \log K + K - 2.$$

Table 1 shows the relationship between T and K for $M = 512$.

K	1	2	4	8	16	32	64	128	256	512
T	511	257	132	73	50	51	76	137	264	520

Table 1 Relationship between T and K for summing

For this task, the optimal partition size is 16.

Example 2: Sorting M numbers

Assume that the task uses the parallel sorting algorithm given in [4] (which is designed for a ring interconnection network). The algorithm takes $(M/K)\log(M/K) + 2M$ comparisons and $2M$ communication operations. Hence, we have

$$T = \frac{M}{K}\log\frac{M}{K} + 4M.$$

Table 2 shows the relationship between T and K for $M = 512$.

K	1	2	4	8	16	32	64	128	256	512
T	6656	4096	2944	2432	2208	2112	2072	2056	2050	2048

Table 2 Relationship between T and K for sorting

For this task, the optimal partition size is $M = 512$.

From these two examples, we can see that the optimal partition size of a task depends on the computation and communication requirements of the task. In the case of summing, the amount of computation required is much less than that of sorting, and hence, the optimal partition size is also smaller than that of sorting. In general, the optimal partition size for a task is determined by its inherent computation and the communication structure as well as the computation and communication capability and organization of the system. This optimal size can vary substantially among tasks.

Impact of Workload Characteristics on Optimal Partition Size

Example 3: Workload consisting of 8 sorting tasks
Assume that each task has size $M = 512$. Refering to Example 2, if we assume the partition size for each task is 512, which is the optimal partition size for a single sorting task, then these eight tasks

have to be processed sequentially. Hence, the completion time for these tasks becomes $T = 8 \times 2048 = 16384$. On the other hand, if we use a partition size of 64 for each task, all the tasks can be processed in parallel. For this case, the completion time for these tasks become $T = 2072$.

Example 4: Workload consisting of a summing task and a sorting task

Assume that both tasks have size $M = 512$. In this case, choosing a partition size of 256 for the sorting task and a partition size of 16 for the summing task and executing the two tasks in parallel, results in the shortest completion time.

From Examples 3 and 4, we can see that the optimal partition size for a task may not be the optimal partition size for the same task when it is part of the workload with other tasks. Hence, while a programmer can analyze the characteristics of the task and determine the optimal partition size for this task, this size may no longer be optimal when the task is being executed. A similar situation occurs if a compiler is used to determine the partition sizes. The dynamic characteristics, which include both workload characteristics and system characteristics (such as the number of available PEs), must be taken into account in order to effectively optimize the performance of the system.

Use of Application Characteristics in the Operating System

In our proposed partitionable system, we assume that the tasks characteristics are available to the system. We have designed an efficient approximation algorithm that, given the workload characteristics (that is, the tasks to be executed), it can determine the partition sizes for the tasks and construct a schedule to executed these tasks. Using this approximation algorithm, the operating system can determine the partition sizes dynamically. While the problem of determine the optimal sizes and schedule is NP-complete, our proposed approximation algorithm can determine an efficient solution, and give the optimal one under certain situation. The details of these results are given in [15].

Summary

In the above discussion, we have only considered the partitioning of the computing resources (that is, PEs). As we have mentioned earlier, the optimal partition size for a task is determined by the computation and communication structure of the task, as well as the computation and communication organization of the system. The communication resource of a system is another critical resource. Our current research in special-purpose partitionable system for robotic appliations include the design of flexible interconnection network in which the bandwidth can be partitioned among different partitions according to the workload and tasks characteristics. Our target design will be a flexible reconfigurable and partitionable computer in which various resources will be allocated dynamically according to the workload and tasks characteristics. The system will make use of the pre-analyzed application characteristics in determining the optimal partitioning of system resources.

CONCLUSION

In this chapter, we have presented our preliminary research results in the REPLCIA project. Our efforts are towards a design methodology, begining with capturing the application characteristics and then deriving the desirable architectural features and required operating system support. Extensive studies are required in analyzing the application characteristics. These studies include a classification of the problems according to their computation and communication characteristics, analyzing the impact of application characteristics on various design decisions (such as communication resources, computation resources, and storage resources).

The premise of this chapter is that in order to design a high performance computer to support robotic applications, the characteristics of the application must be thoroughly understood and analyzed. The importance of this premise is illustrated by the evaluation of the mesh, ring and multi-stage ICNs for multi-processor computers supporting robotic applications. We further demonstrate the importance of this premise by showing that the knowledge of workload characteristics can enable the system to determine dynamically the partition sizes.

REFERENCES

1. Allen, P., *Object Recognition Using Vision and Touch*, Ph.D. Thesis, University of Pennsylvania, 1985.
2. Bajcsy, R., and Allen, P., *Sensing Strategies*, Proceedings of the 2nd ISRR, Tokyo, Japan, 1984.
3. Batcher, K.E., *Design of a Massively Parallel Processor*, IEEE Trans. on Computers, vol. C-**29**, Sept. 1980, pp. 836–840.
4. Baudet, G., and Stevenson, D., *Optimal Sorting Algorithms for Parallel Computers*, IEEE Trans. on Computers, vol. C-**27**, Jan. 1978, pp. 84–87.
5. Briggs, F.A., Hwang, K., Fu, K.S., and Wah, B.W., *PUMPS — Architecture for Pattern Analysis and Image Database Management*, Proceedings of Computer Architecture for Pattern Analysis and Image Database Management, 1981, pp. 178–187.
6. Briggs, F.A., Fu, K.S., Hwang, K., and Wah, B.W., *PUMPS Architecture for Pattern Analysis and Image Database Management*, IEEE Trans. on Computers, vol. C-**31**, Oct. 1982, pp. 969–982.
7. Kapur, R.N., Premkumar, U.V., and Lipovski, G.J., *Organization of the TRAC Processor-Memory Subsystem*, Proceedings of the National Computer Conference, 1980, pp. 623–629.
8. Kuehn, J.T., and Siegel, H.J., *Simulation Studies of a Parallel Histogramming Algorithm for PASM*, Proceedings of the 7th International Conference on Pattern Recognition, 1984, pp. 646–649.
9. Kuehn, J.T., Siegel, H.J., Tuomenoksa, D.L., and Adams III, G. B., *The Use and Design of PASM*, in *Integrated Technology for Parallel Image Processing*, edited by Levialdi, S., Academic Press, London, 1985, pp. 133–152.
10. Kushner, T., Wu, A.Y., and Rosenfeld, A., *Image Processing on ZMOB*, IEEE Trans. on Computers, vol C-**31**, Oct. 1982.
11. Lee, W.H., and Malek, M., *MOPAC — A Partitionable and Reconfigurable Multicomputer Array*, Proceedings of the International Conference on Parallel Processing, 1983, pp. 506–510.
12. Lint, B., and Agerwala, T., *Communication issues in the design and analysis of parallel algorithms*, IEEE Trans. on Software Engineering, vol. SE-**7**, Mar. 1981, pp. 174–188.
13. Ma, Y.W, and Krishnamurti, R., *REPLICA — A Reconfigurable Partitionable Highly Parallel Computer Architecture for Active Multi-Sensory Perception of 3-Dimensional Objects*, Proceedings of the 1st International Conference on Robotics, 1984.

14. Ma, Y.W., and Narahari, B., *Optimal Mappings Among Interconnection Networks For Performance Evaluation*, Proceedings of the 6th International Conference on Distributed Computing Systems, 1986.
15. Ma, Y.W., and Krishnamurti, R., *Partitionable Architectures for Special Purpose Applications*, Technical Report, Department of Computer and Information Science, University of Pennsylvania, 1986.
16. Nutt, G.J., *Multiprocessor Implementation of a Parallel Processor*, Proceedings of the 4th Annual Symposium on Computer Architecture, 1977, pp. 147–152.
17. Pass, S., *The GRID Parallel Computer System*, in *Image Processing System Architectures*, edited by Kittler, J., and Duff, M.J., Research Studies Press, Letchworth, Hertfordshire, England, 1985, pp. 23–38.
18. Paul, R.P., Durrant-Whyte, H.F., and Mintz, M., *A Robust, Distributed Sensor and Actuation, Robot Control System*, Technical Report, University of Pennsylvania, MS-CIS-86-07, 1986.
19. Premkumar, U.V., Kapur, R., Malek, M., Lipovski, G.J., and Horne, P., *Design and Implementation of the Banyan Interconnection Network in TRAC*, Proceedings of National Computer Conference, 1980, pp. 643–653.
20. Siegel, H.J., Siegel, L.J., Kemmerer, F.C., Mueller Jr. P.T., Smalley Jr. H.E., and Smith, S.D., *PASM: A Partitionable SIMD/MIMD System for Image Processing and Pattern Recognition*, IEEE Trans. on Computers, vol. C-**30**, Dec. 1981, pp. 934–947.
21. Siegel, L.J., Siegel, H.J., and Swain, P.H., *Performance Measures for Evaluating Algorithms for SIMD Machines*, IEEE Trans. on Software Engineering, vol. SE-**8**, July 1982, pp. 319–330.
22. Tuomenoksa, D.L., and Siegel, H.J., *Task Scheduling on the PASM Parallel Processing System*, IEEE Trans. on Software Engineering, vol. SE-**11**, Feb. 1985, pp. 145–157.

Chapter 7

DISTRIBUTED HIERARCHICAL CONTROL ARCHITECTURES FOR AUTOMATION

Timothy L. Johnson

Control Technology Branch
General Electric Corporate Research & Development
P.O. Box 43
Schenectady, NY 12345

INTRODUCTION

Feedback effects in computer integrated manufacturing (CIM) systems have important implications for performance, and are a significant factor in the selection of CIM hardware architectures. This chapter provides a first introduction to feedback requirements for CIM systems, and illustrates how a modular hierarchical CIM software architecture, which can satisfy the primary control requirements, is affected by the underlying hardware architecture. The particular implications of manufacturing architectures for robot control in a CIM environment are used to illustrate hardware architecture tradeoffs.

CIM ARCHITECTURE DEFINITION AND REQUIREMENTS

Mid-volume discrete part manufacturing will provide the scope and focus for this discussion, although many of the concepts can be generalized to high-volume or continuous-flow processes. This class of CIM systems is economically important and is representative of the research and target market of many of today's commercial CIM systems. The economics, manufacturing processes, and time scales of systems outside of this scope differ so significantly that in many cases different control software and hardware architectures have more favorable properties than those indicated herein, even though the underlying design concepts are similar. For instance, FMS systems often are designed for high-volume discrete part manufacturing, while DDC systems are usually designed for continuous process control. (1,2)

Fundamental to architecture, generally, is modularity of function, i.e., the ability to represent a design as an interconnection of replicable components. In the context of distributed computing for CIM systems, architecture is more specifically defined as follows: 'Architecture is the choice and configuration of hardware and software modules which constitute the controller design.' This definition encompasses the modularization of both hardware and software functions, as well as their configuration, or interrelationship. This definition is considered to be necessary, due to the technical ability which currently exists to trade off hardware for software functions, and vice versa. "Computer architecture" will be used synonomously with "hardware architecture" herein. Furthermore, a CIM system may be considered to include not only application modules associated with real time control, but also other software and hardware modules associated with materials requirements, financial data, orders tracking, and reporting functions; and additional modules associated with the operating system, user interface, file system, etc. Hardware and software resources, in general, may be shared between control and non control-related modules.

From an industrial perspective, CIM architecture choices are commercially important. The modularization of the problem itself has a profound effect on the versatility, maintainability and extensibility of the resulting product. Production costs and market appeal are also impacted directly by architectural choices. The system configuration, or interconnection of modules, can substantially affect design time, integration cost, and system-wide performance or maintainability. Since different manufacturers operate under very different financial resource, market, and hardware availability constraints, no single architecture is likely to predominate in the near future. The hierarchical software architecture used for illustration in the sequel, for instance, is shown to be suitable for certain control requirements, but it may not be an optimum architecture under conditions of severe hardware resource constraints.

CIM control system architectures are thus regarded as a subset of the collection of software and hardware modules which constitute the full CIM system. In some existing designs, real-time control performance is not a dominant criterion for modularization, and hence control functions may appear to be scattered more or less sporadically over a large set of modules with other primary functions; other existing designs are one-of-a-kind and do not even reflect modularization. These situations are considered to be consequences of the immature current state of the art and are expected to become less common in the future.

Control architecture modules may be distinguished from other modules by their role in providing feedback from command and sensing elements to actuating elements of the system, either directly, or indirectly, through interactions with modules that are directly connected to the manufacturing process. In fact, the existence and interaction of modules with both actuation and sensing functions is one of the features which distinguishes automated manufacturing systems from previous generation manufacturing systems: traditional manufacturing discipline enforced a rigid separation between hardware and software concerned with actuation (usually implemented as a point-to-point interconnection based on programmable controllers) from that concerned with sensing (usually implemented as a separate inspection or management information (MIS) system). The traditional systems preserved (human) job security by forcing all decisions through the human operator. Not only was this practice costly, but it failed to account for the limitations of humans in speed of response and information processing capacity. It did have the advantage of avoiding major errors, but sometimes at the expense of less consistent operation under routine circumstances.

CIM control system architectures may be distinguished from other parts of a CIM system by certain special requirements. Control system architectures usually have a definite purpose: to assure that a production plan is achieved, and to detect and/or correct deviations which occur during the execution of the plan. (In control terms, steady-state tracking and disturbance rejection, respectively). In order to achieve this objective, in most cases, the delay in acting on a planned action, or in responding to a detected event, must be bounded. As the delay in responding to an event increases, the manufacturing process will tend to deviate more and more from its planned course, and eventually full recovery will become difficult or impossible to achieve. Two situations may occur as delay is increased: (1) the actuators lack the power or authority to correct the situation in time, and/or (2) the logic of the recovery process becomes too complex to compute in real time. The maximum tolerable delay depends on the manufacturing process time scales, and on the particular types of sensors and actuators employed. Often, the problem will be made more difficult if the sensors of actuators are widely separated or are poorly chosen, e.g., if an elaborate deduction process is required to detect an important failure, given the sensor measurements available. A further requirement for control system architectures is that they be significantly more reliable than the process which they control. Most production equipment is designed for very long mean-time between failures, and this places a very high requirement on the control system reliability.

These requirements suggest some general properties of CIM control system architectures: In order to be able to execute a plan, the control system must at a minimum be able to issue an appropriately timed sequence of commands to a corresponding set of acuators which are capable of effecting the necessary actions; similarly, the set of measuring devices must be such that both correct and incorrect execution of the commands can be deduced. Typically, additional sensing and acuation elements will be required in order to achieve the correction of errors that arise in off-nominal conditions. At a minimum, the set of actuation and sensing elements must be adequate to maintain the system in a safe state. The structure of both hardware and software modules of a system place constraints on the ability of the control system to provide correctly timed commands to elements of the manufacturing process; while planned actions can be pre-computed in order to compensate for hardware timing constraints, corrective actions must be computed in real time, based on current or recent measurements. Thus, responsiveness to external events must be assured via polling, interrupts, or other means of sampling. Often, continuous-valued (analog) measurements and controls are updated at regular intervals, while set-valued (digital) measurements and controls are updated asynchronously whenever new events are generated or detected. In order to meet requirements of bounded delays in response to measurement changes, the class of computations supported by the architecture normally should have particular properties; e.g., an upper bound on execution time of the algorithms (with respect to different data sets), and a small variability in the execution time. The process layout should avoid the need to rapidly communicate large or highly variable amounts of data between remote locations; this applies not only to the control calculations, but also to any supporting data base elements. In many cases, control and data traffic share the same network elements; the need to frequently transmit large blocks of data increases both the mean and variance of the transmission delay for control information. Finally, the high reliability requirements of CIM control systems imply that the system architecture provide mechanisms for both hardware and software fault-tolerance (integrity).

As an example of these concepts, consider the architecture of a robot controller, as one unit of a CIM control architecture. Some of the requirements of this system unit might include:

1. Workspace dimension: 2m. × 2m. × 2m.
2. Maximum path error: 0.5 cm.
3. Point-to-point maximum path velocity: 5 m/sec.
4. Maximum endpoint error (accuracy or repeatability, or both) 0.1 mm.
5. Endpoint maximum approach/depart velocity: 0.1 m/sec.

Depending on the specific robot geometry, these requirements can be translated into requirements on servo update rates and accuracy. For instance, (2) may require 16 bit accuracy during slewing, while (4) may require 20 bit accuracy for accumulation of the integral of the error near the endpoint. Similarly, (3) may require a 1 msec. servo update rate in order assure a safe emergency stop at maximum velocity, while (5) may require only a 2 msec. update rate to maintain stability. These requirements may imply that high speed bit-serial communication between the axes will be too expensive to meet other cost goals for the system. They may place a bound on the number of multiply operations which can be tolerated in the servocontrol calculation.

HIERARCHICAL ARCHITECTURES FOR CONTROL

Hierarchical control architectures may be regarded as a subset of a broader class of controller architectures termed 'decision in the loop' systems [3], as illustrated in Figure 1. These are systems characterized by a set of decision rules which operate iteratively on a data set consisting of the system goals, the current measurement values or new events, and a set of internal data representing the state of the decision process, and produce at each iteration actuation values for the controlled process and new values of the internal data. This more general class of rule-based system need not involve any particular form or fixed number of rules, and it need not incorporate any particular discipline or order of computation for the rules. Certain production rule and blackboard systems used in artificial intelligence research are of this class. While this class of control concepts can actually be implemented, and while it offers the potential for performance improvements over the more restricted hierarchical architectures discussed here, no engineering design methodology or prior performance analysis tools have yet been developed for it. Presently, the run-time performance and behavior of such systems remains quite unpredictable, and the degree of performance improvement over hierarchical architectures with fixed structures is questionable for today's manufacturing process architectures.

The concept of hierarchy may be applied in many different ways, and the architecture described here is only one of several types which might be termed 'hierarchical.' The notions are essentially

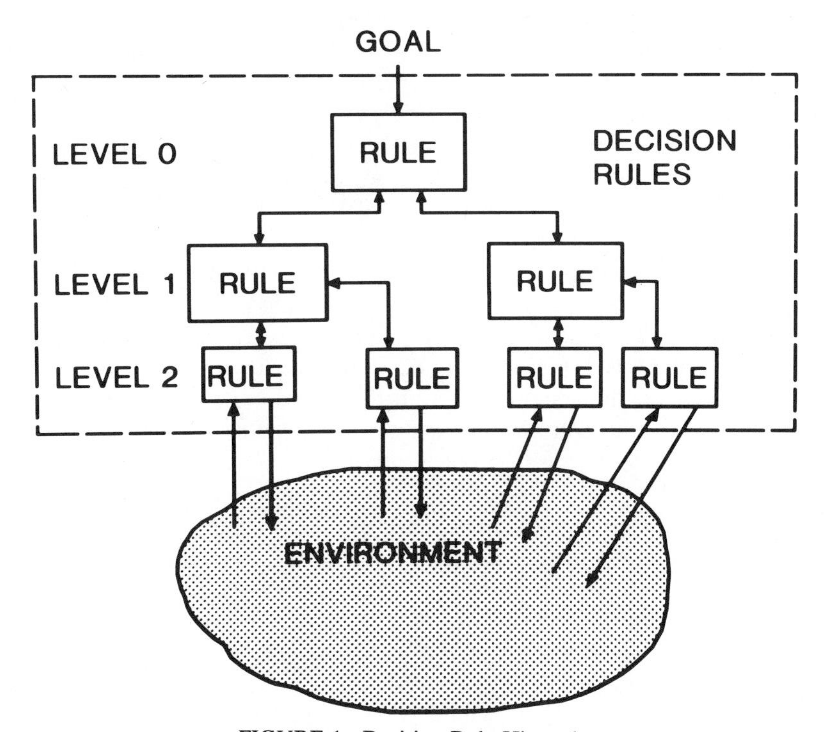

FIGURE 1 Decision Rule Hierarchy

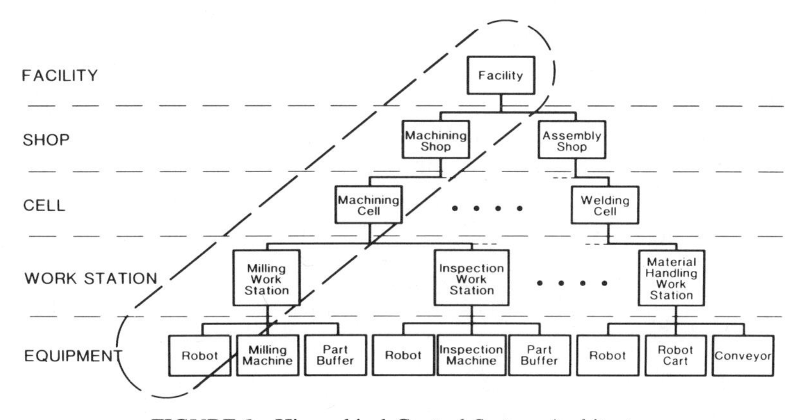

FIGURE 2 Hierarchical Control System Architecture

those introduced by Albus [4] and co-workers, which have been implemented over the last decade at the U. S. National Bureau of Standards Automated Manufaturing Research Facility as illustrated in Figure 2. Various industrial systems exhibit similar structural elements. The software modules of these hierarchical systems each implement a collection of decision rules of the form:

IF (situation) AND (event), THEN (actions) AND (new situation).

The 'situation' or state or mode of the module often reflects an estimate or prediction of what the manufacturing process is doing or is supposed to be doing. The 'event' reflects new information in the form of commands from higher nodes, and new sensory information passed up from lower nodes or from measuring devices on the process itself. The 'actions' represent commands to lower nodes or sensory information passed up to higher nodes. The new situation is a projection of the next expected situation, assuming that the command is correctly executed. The number of possible situations and events is assumed to be finite (though possibly large). For a given situation, the decision rule imposes a causal relationship between the current event and the subsequent action; if this mapping is restricted to the subset of events consisting of command input changes from higher modules, and the subset of actions consisting of command outputs to lower modules or to the manufacturing process actuators, then it is still causal, and it is the precedence relation defined by this causal mapping which defines the position of a module in the hierarchy. The command to the top (or root) module of the system is supplied by elements external to the control logic. Any terminal (or leaf) module of the hierarchy will have an interface to the manufacturing process through measurement and/or control devices. Modules may exchange information or update a common data base in addition to the control functions indicated above. In summary, the software modules of this class of CIM control architectures may be ordered in a hierarchy on the basis of causality.

Each software module is associated with a particular computational hardware element; in general several modules will be associated with one element. A given hierarchy of software modules may be associated with many possible hardware architectures, the only constraint being that the interconnection of hardware elements will permit the information exchange implied by the interconnection of data paths between the software modules. For instance, a hierarchical software architecture could be supported on a ring network hardware architecture. Clearly, not all hardware architectures will provide the same real-time performance, however. If a hierarchical software architecture is partitioned such that every subset is contiguous, and if each subset is associated with a particular processor, then the resulting processor network will be realizable as a tree also. Although this network will support the necessary connectivity of software modules, it may not be optimum because the background data flow, e.g. between modules and the data base, may have a different pattern than the flow of command and sensory information. Thus, it is neither necessary nor appropriate to restrict the processor configuration corresponding to a hierarchical software architecture to be hierarchical. The fallacy that hierarchical software requires hierarchical hardware has often led to incorrect assertions concerning the limitations of hierarchical control. For purposes of this discussion, any network architecture which provides the necessary data paths to support the hierarchical software organization will be considered to be 'compatible' with hierarchical control.

Up to this point of the discussion, we have attempted to provide an answer to the question: 'When is a CIM control architecture hierarchical?' But from an engineering viewpoint, we are more interested in the question of whether hierarchical control architectures can meet CIM requirements, and if so, how to design them. To answer these questions, we shall first outline a design methodology, and then show how this methodology can produce systems which meet the major control requirements.

Almost any manufacturing task may be associated with the transformation of raw materials or subassemblies into parts or final products. In most cases the transformation can be accomplished through one or more ordered sequences of operations. Each operation, in turn, may be associated with a number of more detailed steps or substeps. Each operation is also associated with a set of necessary equipment, personnel, time, and materials which are consumed as inputs and/or removed as scrap. A process plan maps out the necessary sequence of operations and resources required to produce each part or product to be manufactured by a CIM system. For purposes of this discussion, assume that the necessary resources have been determined, and that a layout of production equipment capable of meeting the process resource requirements has been specified. Usually, the operation sequence required to produce each part or product can be organized hierarchically; completion of a sequence of steps at a lower level is associated with completion with a single higher level operation; a sequence of higher level operations may constitute a production stage. etc. The nodes in a control hierarchy are placed in correspondence to the functional decomposition of the manufacturing process, e.g., as expressed in the process plan.

The purpose of each node is to manage the planning, performance, and/or recording of the task with which it is associated. The key feature of the node is performance; planning and reporting are sometimes centralized for convenience of implementation. The basic requirements to assure performance of a task are (a) to interpret the higher-level command from the next higher task entity, or from the operator, into an appropriate sequence of sub-commands (a simple form of planning); (b) to reserve the necessary productive and computational resources to carry out the task; (c) to actuate the necessary lower-level nodes, or the process itself, to complete the necessary sequence of operations, (d) to monitor the results of the actions and to determine normal or abnormal completion of the task, (e) to acknowledge completion of the task or error conditions to the next higher task entity, and (f) to release resources when the task is complete. Typically, a nominal sequence of events will be required for successful completion of a task; by design, the factory should be constructed to maximize the probability of normal completion. If an abnormal event is detected, the resources associated with the task may or may not be sufficient to correct it; if they are, corrective action (involving feedback) may be taken — if not, then the event is normally reported to a higher node which commands more resources, and the local resources are placed into a safe state. More advanced systems are associated with more extensive planning, resource management, error detection and correction, and knowledge base manipulation than simple systems.

Each node performs the functions of a feedback control subsystem. Its input from the higher node corresponds to a command input, which specifies the desired action; its input from a sensor or from a lower node provides measurement information which can be used to correct the nominal action according to the actual state of the manufacturing process; and its outputs serve to actuate the plant and to communicate its actions to other parts of the control system. In virtually all cases, the internal logic of the node may be organized into a sequential state machine with states corresponding to estimated conditions of the process segment which the node is controlling. A representation of the state of the controlled process may require access to and/or updates of a manufacturing data base, as indicated in Figure 3. The functions of command extrapolation. (a form of feed-forward control), state estimation, and error correction are familiar to control designers. The functions of error detection, planning, and data base update have been treated in more advanced works on automata theory [5], decision theory [6], and stochastic control [7].

In fact, the entire hierarchical control system, consisting of the graph and the node functions on it, is a particular type of decentralized control system. The lowest level (leaf) nodes, being connected by sensors/and or actuators to the manufacturing process, function as the inner loops of a control system. The higher levels function as successive layers of 'outer loops,' reflecting a classical design philosophy. Two aspects of the classical methodology provide important insights into successful design of hierarchical manufacturing systems.

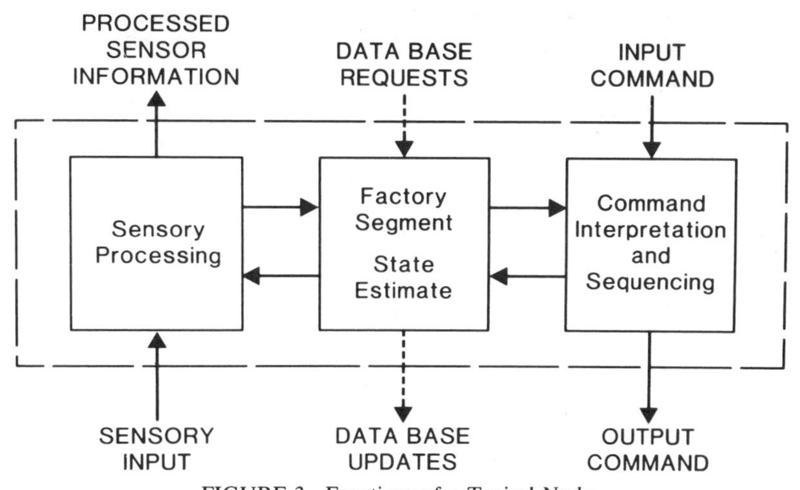

FIGURE 3 Functions of a Typical Node

First, the inner loops are designed to exhibit the shortest closed loop time scales, resulting in the outermost loops having the longest time scales. Secondly, the loop transient response characteristics are designed to become more regular and better-behaved in progressing from inner to outer loops. The nasty properties of the raw process are corrected by successive layers of the design, resulting ultimately in a system which is more robust to failure and more predictable. Traditional inner loop/outer loop analysis validates these properties when the design guidelines are followed; they are not an inherent property of any inner/outer loop (or hierarchical) system!

Hierarchical control architectures have a number of other interesting features. Since the hierarchy is specified by a tree connectivity graph, each node may be specified in terms of only the local variables which designate its single parent node and a finite number of offspring nodes; this locality of reference simplifies the generation of new nodes. If data base updates are time slice synchronized, so that all nodes read (in any order) before all nodes write (in any order), then a hierarchy automatically avoids read-read, read-write, and write-write conflicts. This property is not true for other software architectures such as networks of modules. Another property of a hierarchy is that the number of inter node linkages decreases semimonotonically at each layer; thus, there may be a single (bidirectional) link to the apex node, but a large number of links (one for each sensor and/or actuator interface to the manufacturing process) at the base level. Taking a cross section at a particular level of the hierarchy, the lower layers of the controller, connected to the manufacturing process, can be viewed as a 'partially controlled' plant; the collection of state estimates used in the controller nodes just above this layer constitute an abstracted or simplified view of the partially controlled plant. This abstraction principle is another useful design concept that can be applied to hierarchical controllers; it can be stated in rigorous mathematical terms for certain classes of systems [8].

In the process of designing hierarchical control systems, users have discovered additional design practices which are useful: (a) A task may admit many nearly equivalent decompositions. Although the general principles governing task decomposition are not yet very well understood, well balanced designs are known to avoid bottlenecks. In a well balanced design, tasks at the same hierarchical level should share the same degree of complexity, similar completion times, comparable degrees of uncertainty, comparable and non-overlapping resource requirements, and good potential for concurrency. A candidate task decomposition which doesn't meet these criteria can sometimes be balanced by stratifying the tasks at different levels (e.g., by adding levels). (b) Errors should be identified and resolved at the lowest level possible; an error which is resolved at too high a level will require that the higher node create an additional estimate of the error process at the lower node, which will often be less accurate than the local error estimate, and will often be delayed. (c) Several methods are available for maintaining responsiveness in a hierarchy: use of different sampling rates at different levels, polling, use of cyclic state subsets, use of an 'alert' variable which signals any one of several error conditions. Systems which have been implemented tend to fall between the extremes of time-slice synchronized shared memory systems, such as that at the U.S. National Bureau of Standards, and object-oriented message-passing systems which are readily supported by a number of multi-processing real-time operating systems. For typical manufacturing processes, the response and communication times between nodes are usually 10-100 times as fast as the process time scales; thus, in the occasional situation where information must travel up and then down a hierarchy, the delay is not usually significant. If such delays become significant, or if the amount of this type of traffic is large, the hierarchy has been poorly designed. (d) Resource allocation in a hierarchical control architecture often admits a dual approach: at a high level, a plan is generated which governs the scheduling of new jobs into the system. At lower levels, deviations from the plan may be resolved by simple first come, first served methods, priorities based on expected completion times, or other relatively in uncomplicated rules. The high level plan prevents the formation of bottlenecks, which are precisely those situations where the simple-minded rules do not perform well! Traditional rigidly synchronized long-term scheduling algorithms are not particularly well suited to hierarchical control; an appropriate algorithm is one which permits dynamic rescheduling whenever an error of sufficient magnitude occurs. The default scheduling interval should be matched to the process time scale which is characteristic of the level where scheduling occurs.

In summary, hierarchical control architectures have been shown to meet the major requirements for the CIM application, provided that they are appropriately designed. The clearly defined and relatively long-term goals of a CIM system, as well as the drive for high-reliability equipment, lead to a task decomposition which is naturally hierarchical, and for which errors are the exception rather than the rule. These factors permit a highly efficient CIM control system to be designed in most cases, viz., provided that the design of the production equipment and non-real time software functions are generally well-conceived. In a well-designed system, the responsiveness at a particular level will be well matched with the delays which are characteristic of that level. There is considerable flexibility in the hierarchical design concept to accomodate different process layouts and sensor/actuator configurations. Finally, a hierarchical approach to error resolution provides a robustness which is well-matched to the high reliability requirements of a CIM system.

ROBOT CONTROL IN A CIM SYSTEM

In most hierarchically-controlled CIM systems, the robot controller is associated with the lowest level or levels of the hierarchy. Depending on the degree of refinement of the hierarchy, the lowest level may be designated as the 'work station', 'device controller', or 'joint servo'. In any of these cases, the computational module(s) associated with the controller may be grouped together, so that one may consider the data and control signals which interface the robot to: (a) modules at higher levels; (b) modules at the same level, and (c) modules at lower levels, or electromechanical actuators and sensors, such as motor drives or encoders. The signals corresponding to the first group, (a), usually are transmitted on a single serial link, parallel link, or network link. The traffic on this link will affect the performance of the robot as an element of a CIM system. Consider first the error-free case of perfect preprogramming. In this case, robot programs are sometimes downloaded from the host, having been previously taught and stored, or having been generated from an offline data base. (Note that due to dimensional tolerances, this second possibility is rarely available today; even if it is, the program usually has to be tested on the target machine prior to execution.) In many installations, only a finite number of programs is required, and they are permanently stored in the robot controller; in other cases, programs may be exchanged if local memory is inadequate, or if a higher level module performs minor editing of a master program file. In an idealized error-free situation, the only additional run-time signals on the link will be those used to start and stop a program, to log status information, and to down-load new program files or upload debugged files based on taught path information. If file transfers are only performed when the robot is in a safe state, then at run-time the link is free for only command and status data. In some situations, though, a file may be uploaded or downloaded in 'background' mode while command and status traffic operate in a priveleged "foreground" mode. For instance, a new program file could be downloaded while a resident program is executing concurrently, so that there is no delay in beginning the new program. In this case, the design of the link capacity and protocol must accomodate both sorts of traffic. In the realistic case, unscheduled events may occur in addition to the preplanned, scheduled traffic of the idealized case. Traffic of associated with signals of class (a), for instance, may be augmented by unscheduled starts, steps or tests commanded by higher level nodes, or by unscheduled errors which are detected by the robot software (e.g., a motor failure). In some cases, traffic in both directions may share the same physical link. Unscheduled traffic, while it normally does not involve large data transfers, is often associated with critical response times. In designing the communication link and data protocols, the frequency and distribution of message types and lengths should be taken into account. Often, a worst-case scenario must be considered, forcing the link design to be very conservative. In a rigorously designed hierarchical control system, data of type (b), between processes on the same hierarchical level, should be purely data traffic, and should not contain task commands or acknowledgements. Violation of this principle constitutes subversion of authority in a hierarchy, and generally results in a greatly increased possibility of error, along with an inability of higher levels to monitor what is happening. Links of this type are common in point-to-point networks of programmable controllers, and between robots and equipment which must be synchronized with them. Signals of type (c) are essentially analogous to signals of type (a), but one or more levels lower in the hierarchy. At the lower levels, the transmission protocols are often very rigid (e.g., sampled data for drive commands, phase-modulated pulse trains for encoders), with separate dedicated parallel channels being used for unscheduled events (e.g., power fail interrupts).

In considering aspects of robot control that are specific to robots incorporated in CIM networks, we focus on the link carrying signals of class (a), in the preceding paragraph, between a higher level workstation host and a robot controller. At the host end of the link, the software must usually maintain some partial estimate of the state of the robot; this is the hierarchical version of the 'internal model principle' of control system design. The design of this status acquisition layer must allow for the consequences of discrepancy and/or delay between the host's estimate of the robot state and the actual robot state. A second issue concerns the level of autonomy which is appropriate for the processing capability of both the host and the robot controller. If the robot controller is very primitive (and most inexpensive robot controllers are), then the host software must perform more error handling, replanning, and command interpretation functions. If the workstation host also has limited computing capacity, or is subject to fast response requirements or heavy data traffic, then these functions generally will require more preplanning at higher levels of the hierarchy. A third function of the workstation host is to synchronize and schedule its subservient nodes; this may require special algorithms for status acquisition and maintenance (e.g., a device driver or server), for guaranteeing responsiveness to unscheduled events, and for rapid decision making to minimize the delay in responding on one downlink to an unscheduled event which has occured on another downlink. In the robot controller, at the other end of the CIM link, some method of run-time interrupt management is usually necessary. The controller usually must maintain a status block for periodic transmission to the host. It must have a means of maintaining the robot in a recoverable state in the event of unscheduled new commands from the workstation host, such as 'abort part', or 'stop'. These demands sometimes exceed those which are necessary for a stand-alone robot controller. In some cases, robot controllers will support special channels (discete I/O) for high-speed synchronization with other devices; this does not necessarily subvert authority, provided that both devices communicate status to the host in a timely manner, or if the device which is synchronized to the robot controller is actually slaved to it as a subservient node.

Hardware architecture trade-offs have not been fully considered in this discussion. As a practical matter, hardware above the device controller level in most CIM architectures is vendor-supplied off-the-shelf general purpose computing technology; thus, architectural decisions focus on the choice of vendor(s) and subsequently are constrained by the configuration alternatives supported by the selected manufacturer(s). Key variables in such decisions include the processor cost/speed tradeoffs, the operating system and software development environment, the flexibility of I/O and networking, etc. The implication of these decisions for the robot controller is that it must support one or more standard network or serial link connection to higher-level devices to support signals of type (a). A current architecture dilemma common to many robot manufacturers is how to support signals to type (b), particularly as they may arise from "smart" sensing and actuation devices associated specifically with the robot, such as imaging equipment, force/torque sensors, tool changers and part feeders. Two options are indicated in Figure 4. In the option (a), synchronizing signals from a machine tool and camera controller are brought back to the workstation controller and re-distributed to the robot controller; this configuration is often required today due to the poverty of the serial link (e.g., a single RS-232 link) and the closed architecture provided by many vendors. Evidently, this interconnection does not support rapid intercommunication between the devices attached to the workstation. But several forms of software compensation are still possible. For instance, the robot controller (RC) may have to be designed to ensure safe operation for longer time periods than would otherwise be required, in the presence of misaligned parts; the camera controller (CC) may have to provide a more highly processed image to the workstation (e.g. locations of key part features) to facilitate and expedite the decision process at the workstation level. Another option at the workstation level is to provide a concurrent communication task which intercepts cross traffic (of type (b) above) and passes it directly through between the devices connected to the workstation controller (WS), without any explicit decision making.

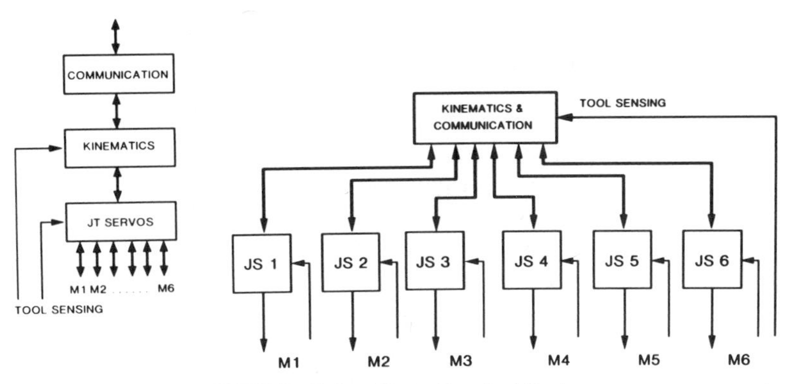

FIGURE 5 Robot Controller Architectures
Key: M = Motor
JS = Joint Servo

properties of this configuration may be higher than the hierarchical network configuration. However, both configuratons will support hierarchical control.

Hardware architecture design within the robot controller is at the component or broad level, and may involve custom VLSI or co-processing chips. Two alternate architectures sketched in Figure 5 illustrate a tradeoff which is common in current controller architectures. The first configuration shows separate processing elements (e.g. boards within the controller) for communications, kinematics, and motion control. Note that the joint sensing the tool sensing enter at different levels. At the motion control (joint servo) level, this architecture might require a single high-performance processor/coprocessor configuration to serve all of the joint servo functions, such as position loop closing and limit protection. The second configuration indicates a common processing element for kinematics and communiction, but separate processors to serve each joint; in this case, joint position sensors are interfaced to individual servo boards. Tool sensing, which may be done in a coordinate frame that involves the positions of several robot axes, provides information at the kinematic level. Again, both of these architectural alternatives are compatible with a hierarchical software decomposition; however, they correspond to different task decompositions. They will have different performances, degrees of reliability, and costs.

CONCLUSION

In this chapter, manufacturing control system architectures have been defined at the conceptual level, as a collection of modules having both hardware and software attributes and interconnection rules. Control architecture requirements for computer integrated manufacturing (CIM) systems have been defined. The particular class of modular hierarchical control architectures has been defined in the context of CIM applications, and has been shown to be capable of meeting most of the CIM requirements, providing that certain design practices are used. Finally, the implications of the CIM requirements for the design of robot controllers, as elements of CIM networks, have been examined. The reader is referred to [9]–[19] for further topics of interest.

Computer integrated manufacturing, as well as robot control, must still be regarded as being at an early stage of development. A large number of new research issues are emerging, and design theory and methodology at still in their infancy. In a rigorous sense, it is not yet possible to 'prove' that hierarchical control systems arise naturally as the solution to particular optimal control problems. However, in a qualitative sense they can be shown to satisfy a large number of important control requirements. A key contribution of this chapter has been to show that there is also a close association between hierarchical control system architectures as practiced in manufacturing, and control system design practices which have developed in other fields and come to be accepted as routine principles in the design of continuous-state control system.

Since control elements are only one component of CIM architectures, they are not necessarily the most significant factor influencing the progress in factory automation. For instance, the evolution of standards for factory communications, such as MAP, is having a profound influence on product architecture decisions. In the area of robot control, different trends are apparent in Japanese and U.S. markets: the Japanese market appears to favor less expensive, less intelligent robot controls, with more functions being performed at higher levels, while the U.S. market seems to favor more local sensor feedback and error-handling, leading to more expensive but more versatile robots. Considerable economic importance may be associated with the evaluation of these alternative development trends.

REFERENCES

1. Kimemia, J.G. and S.B. Gershwin (1983), *An Algorithm for the Computer Control of Production in Flexible Manufacturing Systems*, *IEEE Transactions*, Volume **15**, No. 4, December 1985, pp. 353–362.
2. Harrison, T.J. (ed), *Distributed Computer Control Systems*, Pergamon Press, Oxford UK, 1980.
3. Johnson, T.L., *Hierarchical Decision-in-the-Loop Processes*, *Proc. Oakland University Conference on Aritifical Intelligence*, Rochester, MI, April 26–27, 1983.
4. Albus, James S., Barbera, Anthony J., Nagel, Roger N., *Theory and Practice of Hierarchical Control*, *Proceedings of the 23rd IEEE Computer Society International Conference*, September 1981.
5. Kalman, R.E., Falb, P.L and Arbib, M.A., *Topics in Mathematical System Theory*, McGraw-Hill, New York, NY 1969.
6. Bellman, R., *Dynamic Programming*, Princeton University Press, Princeton, NJ, 1957.
7. Bar-Shalom, Y. and E. Tse, *Dual Effect, Certainity Equivalence and Separation and Stochstic Control*, *IEEE Transaction on Automatic Control*, Volume AC19, pp. 494–500, 1974.
8. Johnson, T.L., *Design of Hierarchical Control Systems for Automated Job Shops*, *Proc. 1984 IEEE ACC Conf.* San Diego, CA, June 1984, pp. 1062–1064.
9. Albus, J.S., McLean, C.R., Barbera, A.J., Fitzgerald, M.L., *An Architecture for Real-Time Sensory-Interactive Control of Robots in an Automated Manufacturing Facility*, *4th International Federation of Automatic Control Symposium*, Gaithersburg, MD, October 1982.
10. Albus, James S., Barbera, Anthony J., Bloom, Howard, Fitzerald, M.L., Kent, Ernest, McLean, Charles, *Hierarchical Control of Robots in Automated Factory*, *Proceedings of the 13th International Symposium on Industrial Robot/Robots 7*, Chicago, IL, April 19, 1983.
11. AlDabass, D. and Rutherford, D., *Simulation Techniques for Microprocessor-based Parallel Architecture*, *Parallel Computers-Parallel Mathematics*, North-Holland, 1977, pp. 125–131.
12. Barbera, Anthony J., Fitzerald, M.L., Albus, J.S., *Concepts for a Real-Time Sensory-Interactive Control System Architecture*, *Proceedings of 14th Southeastern Symposium on System Theory*, April 1982.
13. Bard, Y., *Some Extensions to Multiclass Queueing Network Analysis*, in *Performance of Computer Systems*, (M. Arato, ed.), North Holland, Amsterdam, 1979.
14. Cassandras, C., *A Hierarchical Routing Control Scheme for Material Handling Systems*, *Proc. First ORSA/TIMS Conf., on FMS*, August 1984.

15. Gershwin, S.B., Hildebrant, R.R., Suri, R., and Mitter, S.K., *A Control Theorist's Perspective on Recent Trends in Manufacturing Systems*, *Proc. 23rd IEEE CDC Conf*, December 1984, pp. 209–225.
16. Hildebrant, R.R., *Generating and Implementing Schedules for Time-Critical Manufacturing Processes*, *Proc. 23rd IEEE CDC Conf*, December 1984, pp. 236– 240.
17. Johnson, T.L. and Milligan, S.D., *Emulation/Simulation of Hierarchical Control Systems*, *Proc. 21st IEEE Conf. on Decision and Control*, Orlando, FL, December 1982, pp. 360–361.
18. McLean, C. and Mitchell, M., *A Computing Architecture for Small Batch Manufacturing,*, *IEEE Spectrum*, May 1983.
19. Shin, K.G. and Malin, S.B., *A Hierarchically Distributed Robot Control System*, *Conf. Proc. IEEE Computer Society: 4th Annual Int. Computer Software and Application Conf.*, October 1980, pp. 814–820.

Chapter 8

HYPERCUBE ENSEMBLES: AN ARCHITECTURE FOR INTELLIGENT ROBOTS

Jacob Barhen

Center for Engineering Systems Advanced Research
Oak Ridge National Laboratory
P.O. Box X, Oak Ridge, TN 37831

The advent of VLSI technology, and its implementation in the design of a new generation of computer systems involving a high degree of concurrency, is opening tremendous opportunities for a wide spectrum of applications which require 'supercomputing' capabilities. We focus on hypercube ensembles, a revolutionary advance in the expanding domain of concurrent computation machines. Critical issues in the development of algorithms for such message-passing architectures are addressed in the context of intelligent autonomous systems, with emphasis on task scheduling and load balancing. We also present the detailed implementation of the Newton-Euler inverse dynamics equations for a robot manipulator on the NCUBE supersystem.

Keywords: concurrent_computation, hypercube, inverse_dynamics, load_balancing, machine_intelligence, NCUBE, robotics, supercomputers, VLSI

I. INTRODUCTION

The advent of VLSI technology,[1] and its implementation in the design of advanced computer systems involving a high degree of concurrency,[2] are opening tremendous opportunities for a wide spectrum of applications[3] which require 'supercomputing' capabilities. The basic trend is to use state-of-the-art VLSI to integrate an entire processing system on a single chip,[4-6] including communication links, memory interface, 32-bit processors[5-6] and even 64-bit IEEE floating point,[6] resulting in smaller and cheaper processors comparable in performance to their larger and more expensive predecessors. This continuing trend appears to be the major technological drive behind concurrent computation, i.e., the use of an ensemble of small computers that work concurrently on parts of a complex problem, and coordinate their computations entirely by sending messages to each other.

Some of the most challenging computational problems facing scientists and engineers today arise within the framework of intelligent autonomous systems.[7] Such problems range from robots operating in unstructured hazardous environments where explosives, toxic chemicals, or radioactivity may be present, to the development of battle management paradigms for the Strategic Defense Initiative.[8] To enable a robotic system to work effectively in real time in an unstructured environment, one needs to solve repeatedly a variety of highly complex mathematical problems such as on-line planning, vision, sensor fusion, navigation, manipulator dynamics and control. Similarly, a 'Star-Wars' defense system would need to respond to an offensive strike by coordinating perhaps millions of separate actions on a schedule timed in milliseconds.[9] The computational requirements of these problems fall into the 'supercomputer' class, but ultimately we need to solve them 'onboard' the autonomous robot or space system. The only realistic option is VLSI-based concurrent computation. However, in order to achieve the required real-time performance, the new generation of concurrent computers must be provided with the capability of scheduling very efficiently sets of tasks among which complex interrelationships (such as precedence constraints) may exist. A more general requirement for achieving utmost efficiency in the utilization of the concurrent computer is to dynamically balance the computational load among each processor in the system.

This chapter is organized as follows. Section II provides a general background on hypercube ensemble architectures and outlines some of the critical issues involved in the design of algorithms for concurrent computation. The NCUBE supersystem is introduced in Section III. Then we address the fundamental research issues of task scheduling and optimal resource utilization in such a message-passing environment. In Section V we present the detailed implementation of the Newton-Euler inverse dynamics equations for a robot manipulator on the NCUBE machine. We conclude by summarizing the main points of this paper.

II. HYPERCUBE ENSEMBLES: ARCHITECTURAL FEATURES AND CRITICAL ISSUES IN ALGORITHM DESIGN

Several types of parallel computer architectures involving different tradeoffs are available for the design of intelligent machines. One of the most fundamental choices is between single instruction-multiple data (SIMD) stream and multiple instruction-multiple data (MIMD) stream multiprocessors. The former, in which a single instruction controls all of the processors while they operate on different data, are simpler. MIMD machines, where each processor is independent of the others and executes its own program, are more general-purpose.

The second major design tradeoff is between shared and local memory. In a shared memory system, processors can access either a single large memory bank using a bus, or a distributed memory using special switching hardware (the 'stunt' box). In a local memory system, each processor can only access its own (local) memory. A shared memory computer is hard to scale up to a large number of processors since either the cost and complexity of the switching hardware grow much faster than the size of the machine, or the bus to memory saturates and becomes a bottleneck above a few processors. A breakthrough in this area in still within the realm of research.[10] The local memory (or ensemble) architecture has the important advantage of not being inherently limited by the number of processors that can be added to a concurrent computer system. However, these processors can communicate only through message passing.

II.1. Hypercube Ensembles

By 'hypercube ensemble machine' we refer to a MIMD local memory multiprocessor design in which $N=2^d$ identical microprocessor 'nodes' are connected in a binary d-dimensional cube topology using fully asynchronous bidirectional channels.[11] For illustrative purposes, a few hypercubes of low order are shown in Fig. 1, where circles denote nodes and lines refer to communication channels. It is important to notice that hypercubes can be constructed in a modular fashion, i.e., an order-d hypercube is constructed from two order-(d-1) cubes by connecting duplicate nodes.

Several architectural characteristics make the hypercube machine attractive for real-time robotic applications. The first refers to communication time between nodes versus wiring complexity (cost). For example, consider a 12-dimensional cube, i.e., $N=2^{12}$ processors. This cube is comparable to a 64×64 square grid, a $16 \times 16 \times 16$ cubic grid, a 12-level deep binary tree or a 4096 node ring. The 'cheapest' configurations are the binary tree (8190 links) and the ring (8192 links). However the former is subject to single point failure, while the latter requires 2049 'hops' between the most distant nodes; this distance is reduced to 126 hops for the square grid and only to 12 hops in the hypercube. For communication speed the optimal choice would obviously be a fully connected configuration, but this is unacceptable since it requires close to 17,000,000 links! The hypercube appears the best design tradeoff.

Second, the hypercube is the natural network for one of the most important of all computational algorithms, the Fast Fourier Transform (FFT), which is used in all areas of scientific computation, and in machine intelligence applications such as sensor signal processing.

Third, the hypercube is a superset of most of the other candidate networks. For example, a grid of any dimension can be mapped onto a hypercube by 'ignoring' an appropriate set of hypercube connections.

Fourth, the hypercube is densely enough connected that a viable approximation is to consider that every node is connected to all other nodes. This is important since in some machine intelligence computations, the pattern of communication is not predictable. It would obviously be better in that case that all nodes have actual

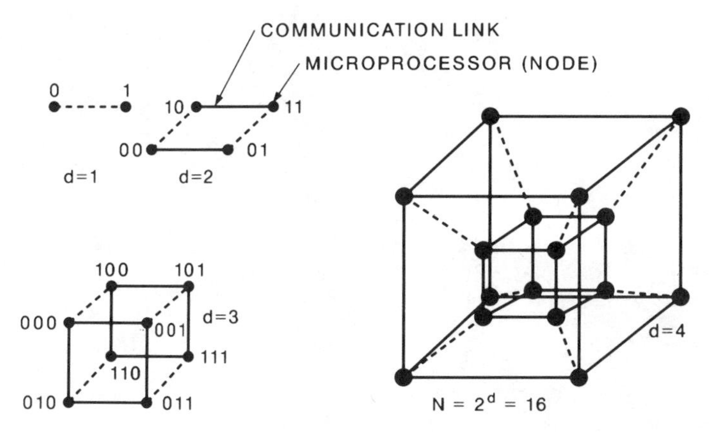

FIGURE 1 Hypercube Architecture in d Dimensions.

connections to all others, but, as pointed out in our example above, that is unwieldly for a large network. Thus, a complete connection network lacks the fifth property of the hypercube: It is the most densely connected network that is also scalable to thousands of processors. For example, in order to double the number of processors in a system, it is only necessary to add one communication channel to each node.

Finally, a hypercube looks topologically identical from the point of view of each node: there are no corner versus edge, or root versus leaf nodes, as there are in regular grids or trees. This symmetry is particularly attractive for simplifying the dynamic reconfiguration of the system.

Such considerations have recently led to the successful development of several families of hypercube ensemble machines. Work at Caltech, for example, ranges in scope from the 'cosmic cube'[11] (initially 64 nodes connected as a 6-cube, using 8086/8087 16-bit processors and currently upgraded to 68020's) to the 'mosaic experiment'[4] (which involves single-chip nodes). In a similar vein, the much finer grain 'connection machine' being developed by Thinking Machines Corp., is reported [12] to have implemented processor-to-processor communication through a fast message routing system that forms a hypercube.

The concurrent computation system currently being investigated

FIGURE 2 NCUBE Concurrent Processor in the CESAR Laboratory for Machine Intelligence.

at the Oak Ridge National Laboratory's Center for Engineering Systems Advanced Research (CESAR) is also based on a hypercube architecture and is shown in Fig. 2. This system, developed by NCUBE Corporation, was conceived from the ground up to be optimally implemented in state-of-the-art VLSI. A 10-dimensional cube is designed to provide (following an initial testing period) potential raw performance of 500 Million Floating point Operations per Second (MFLOPS), or 2000 Million Instructions Per Second (MIPS), yet is extremely compact in size. The initial CESAR implementation is a 6-d cube, which became fully operational in January 1986.

II.2. Developing Algorithms for Ensemble Machines

There are roughly two classes of applications for concurrent computation ensembles. We shall refer to them as Class-I and Class-II in the following discussion. Class-I applications, generally representative of problems encountered in the classical fields of science and engineering, are characterized by a very regular (crystalline) structure in space and time that is known statically; they are computation-intensive, and tend to use the entire machine in the service of a single activity. Although the corresponding formulations (including, for example, matrix, grid or finite element computations) are usually very demanding, their process structure (task graph) is so regular that it can be mapped easily onto the hardware topology. Class-II applications tend to have the opposite characteristics, of being irregular in time and space, with the irregularities sometimes unpredictable in advance, just as often communication-bound as computation-bound, and having to share the ensemble among several activities. Typical of Class-II applications are problems arising within the framework of intelligent autonomous systems. Threat tracking in an SDI environment, for example, involves detecting missiles being fired, tracking them, allocating weapons, determining whether those weapons have hit their targets, and retargeting. This means that optimal (or near-optimal) mapping of the process structure onto the computational ensemble is a much more general and difficult problem for Class-II applications than for Class-I applications, particularly since the mapping must be computed on a dynamic, event-driven basis.

Even for 'static' process structures (i.e., those with a non time-varying topology) where the mapping can be computed prior to execution, optimal mapping is extremely difficult, particularly when precedence constraints (i.e., task T_5 must be finished before tasks T_6 T_{12}, T_{23},... can be started) are involved.[13] A prototype mapping system, ROSES, developed for our DOE robotics activities,[14] is currently being tested, and shows excellent promise for static process structures. An application for which a near-optimal mapping has been achieved is the solution of a robot arm's inverse dynamics equations, which is briefly discussed in Section V.

The situation becomes considerably more complex if the process structure evolves dynamically. This would be the case, for example, in threat tracking and battle management in a 'star wars' environment. In a more generic sense, handling dynamic process structures underlies the operation of intelligent machines in highly unstructured environments. Complications include the development of appropriate methodologies for handling hard real-time constraints (i.e., scheduling tasks with stringent execution deadlines), the capability for processes to spawn or annihilate other processes, and most importantly, for the computer's operating system to be capable of load-balancing the activities of all processors to achieve near-optimal utilization and throughput.

III. THE NCUBE SUPERCOMPUTER

We now present a brief overview of the NCUBE concurrent computation system. Additional details can be found elsewhere.[6–7] The NCUBE is a rather revolutionary new machine that combines a very significant potential raw performance (500 MFLOPS or 2000 MIPS with 1024 nodes; over 1.4 gigabytes per second I/O bandwidth) with significant cost-performance improvement over existing large computer systems. The keys to this major advance are the concepts of concurrent processing and "system-driven" VLSI.

III.1. NCUBE Node Architecture

Each NCUBE node is an independent processor with its own local memory and communication links to other nodes in the system. In most of the parallel systems being proposed today, a node consists of many chips, often more than 100. In contrast, an NCUBE node has only 7 chips and 6 of them are memory (DRAM). All node logic except the memory has been integrated into a single VLSI chip of about 160,000 transistors; it is shown in Fig. 3.

The memory interface implements a 16-bit data path with an Error Correcting Code (ECC) that detects all double bit errors and corrects any single bit error. In the current ORNL implementation a node has 128 Kbytes of physical memory using 6 DRAMs organized as 64K by 4 bits.

The node processor is a general purpose, 32-bit computer. It has a fully symmetric set of arithmetic and logical operations on 8, 16 and 32 bit integers. It also conforms to the IEEE 754 floating point

FIGURE 3 The NCUBE Processor Chip.

standard with operations on both 32 and 64-bit real data (including hardware square root). There are 16 general registers, each 32 bits long, and a set of powerful addressing modes including support for vector and matrix operations.

Each node is connected in a network to a set of its neighbors through Direct Memory Access (DMA) communication links that are controlled by the processor. They allow for both ordinary message sending and more powerful broadcasting in a high performance interrupt driven operating system. The DMA links, once up, run independently of the processor which can continue computation in parallel with the communication. When a link finishes sending or receiving a message, the processor is interrupted, so no polling is required. A node has 22 DMA communication links, 11 in-bound and 11 out-bound; each link is designed to run at 10 Megabits/sec. Two of the links are used for system I/O which will be explained next. The remaining twenty links are used to connect a node to its neighbors in the hypercube network.

III.2. The NCUBE I/O System

The purpose of an I/O system is to transfer programs and data from disks or other devices (e.g., sensors) into the hypercube array and to transfer computed results from the array to output devices such as end effectors. The "NCUBE/ten" configuration, for example, has eight "system I/O channels" for such data transfers. Every channel will potentially move data at 90 Megabytes/sec in each direction. A system I/O channel consists of one pair of communication links from each of 128 nodes bundled together and brought through the backplane to one of the I/O slots. An I/O Board interfaces with a system I/O channel, through 128 pairs of DMA links. This is accomplished by having a set of 16 NCUBE nodes per I/O Board, and using 8 of the 11 pairs of communication links for system I/O. The backplane of the NCUBE/ten has 16 slots for Processor Boards and 8 for I/O Boards. If all slots are occupied, then there are 1024 nodes and each one has a direct connection to an I/O Board through one of the system I/O channels.

At least one of the I/O Boards in an NCUBE system must be a 'Host' Board. This board runs the 'user interface' including the operating system, editors and translators. It also controls the standard peripherals. Each Host Board has Intel 80286/80287 processors with 4 Megabytes of ECC memory. The operating system is 'UNIX style' and has translators for assembly language, Fortran and C; it controls up to four disk drives of up to 500 Megabytes each. If there is more than one Host Board, then an iSBX connector is used for a high speed inter-Board bus. The Host Board also has a number of miscellaneous functions such as a real time clock and temperature sensors for automatic shutdown on overheating.

III.3. Physical Implementation

The NCUBE hypercube is implemented on a set of Processor Boards. Each board is 16″ by 22″ and contains 64 processing nodes (see Fig. 4). The 64 nodes on each board are connected in an order-6 hypercube. When two boards are inserted, the wiring in the backplane creates an order-7 hypercube; when all 16 boards are inserted, an order-10 hypercube with 1024 nodes is obtained. This is the source of the 'ten' in the name 'NCUBE/ten': the maximum order of hypercube available. For entry level autonomous robots such as HERMIES (see Fig. 5) it appears advantageous to consider the NCUBE/four. This is a board with four NCUBE nodes and a PC-AT bus interface, which is software compatible with the NCUBE/ten. Up to four boards can be used in a PC-AT system, with a raw targetted performance of 8 MFLOPS or 30 MIPS. This illustrates the value and importance of 'system driven' VLSI.

FIGURE 4 The NCUBE 64-node Array Board. Up to 16 boards can be included in an NCUBE/ten system.

FIGURE 5 HERMIES-II, the CESAR mobile robot, is currently used for autonomous navigation experiments.

III.4. Basic System Software

The NCUBE operating system has two components. AXIS, a virtual memory, multitasking, multiuser system with a user interface very similar to UNIX, runs on the Host Boards. VERTEX is a nucleus that runs on each of the nodes.

AXIS treats the hypercube array as a 'device' that can be allocated in subcubes. This facility allows users to request cubes of optimum size for given applications. This is essential for robotic applications where several 'activities' (e.g., sensing, planning, dynamics, ...) must be carried on simultaneously. In addition to the unique subcube allocation capability, AXIS provides for loading, running, communicating with and debugging programs in the hypercube nodes. These facilities are implemented in cooperation with VERTEX.

VERTEX is a small operating system nucleus residing on each node that has facilities for message handling and for process loading, scheduling and debugging. VERTEX automatically routes messages through optimal paths to their destinations. Debugging is accomplished by receiving and responding to a request from the user on the Host Board to set a breakpoint, display memory or registers, change memory, etc.

IV. LOAD BALANCING AND TASK SCHEDULING

Load balancing algorithms[15] are required for dealing explicitly with the allocation of resources in a concurrent computation ensemble. The goal is to minimize execution time by evenly distributing the task loads across the system, while minimizing interprocessor communication. The difficulty in solving this problem lies in the conflict of constraints over a configuration space which grows exponentially with the number of tasks. In particular, the goal of minimizing interprocessor communication, to avoid saturation effects which degrade performance, requires that tasks be 'clustered' on few, adjacent nodes; on the other hand, to best utilize the processor resources requires that tasks be spread out over all nodes.

The load balancing problem is closely related to multiprocessor scheduling, a subject matter which has been studied extensively[16] over the past twenty years, and for which excellent reviews can be found in the literature.[17] Major difficulties arise when the number of tasks required by a particular algorithm exceeds the number of available processors, and/or when the interconnection topology of the task graph, as obtained from the precedence constraints, differs from the interconnection topology of the computation ensemble. Optimal schedules are in general extremely difficult, if not impossible obtain, since for an arbitrary number of processors, unequal task processing times and non-trivial precedence constraints the problem is known to be NP-complete.[18]

IV.1. Basic Concepts in Load Balancing

Static load balancing methods permanently assign newly created processes to what appear at that moment to be the best nodes.

These processes are not moved once their execution is initiated, under the assumption that their runtime characteristics do not later change in such a way as to cause nodes to become very unbalanced. Load balancing thus occurs only when a new process is created. For precedence constrained tasks this is the current state of the art.

To adapt to potential changes in the runtime characteristics of processes, one needs to develop "dynamic" load balancing algorithms. Such algorithms may require that processes be migrated during their life-time to better nodes to provide much needed efficiency, particularly for an ensemble that shares multiple activities, e.g., the 'brain' of an intelligent robot. Load balancing would occur at any time, rather than being limited to times when new processes are created.

To address the static load balancing problem, we are currently exploring the applicability of the simulated annealing method.

IV.2. Simulated Annealing

Experiments aimed at studying the properties of a material near the ground states are often carried out using a process referred to as annealing: the system if first heated up to some high temperature ('melting'), then the temperature is slowly reduced until the lowest energy state is reached. At each temperature during the annealing process the system comes to equilibrium. At equilibrium, the probability of the system being in a certain configuration (state) is governed by the canonical[19] distribution at that temperature. If the cooling is too rapid, various defects may become frozen into the structure.

Kirkpatrick et al. have recently pointed out the analogy between the behavior of condensed matter at low temperatures and combinatorial optimization problems. They proposed a new optimization methodology,[20] referred to as 'simulated annealing', which uses techniques suggested by statistical mechanics to find global optima of systems with large numbers of degrees of freedom. The simulated annealing algorithm can be sketched as follows. Consider a combinatorial optimization problem specified by a finite set C of configurations (or states) X, and by an objective function E defined over X. From equilibrium statistical mechanics we know that all configurations $X = (x_1, \ldots, x_N)$ are possible, but that the probability of observing a given X is governed by the canonical distribution:

$$p(X) = \frac{\exp[-E(X)/\theta]}{\sum_{X \in C} \exp[-E(X)/\theta]} \tag{1}$$

Here θ refers to the product kT of the Boltzmann constant by the absolute temperature for a physical system, and will represent a control parameter ['effective temperature'] in the optimization analogue. The problem is then to find the configuration X which induces the minimum value of E.

The algorithm starts from an initial state X and follows a sequence of annealing temperatures $\theta_0, \theta_1, \ldots, \theta_i, \ldots$ where $\theta_{i+1} < \theta_i$. It can be summarized as follows:

[1.0] loop over temperature index i
  [2.0] set $\theta = \theta_i$
  [2.1] loop over sample size at temperature θ
    [3.0] generate new state $X' = F(X)$ where
      F represents a heuristic for selecting new states
    [3.1] $\triangle E = E(X') - E(X)$
    [3.2] If $\triangle E \leqslant 0$ then
        accept new configuration unconditionally
        i.e. $X = X'$
      Else
        accept new configuration only if it satisfies
        the Metropolis Criterion[21] i.e.
        r = uniform_random (0,1)
        If $r < \exp[-\triangle E/\theta]$ then $X = X'$
      End if
  [2.2] end loop over sample size at temperature θ
  [2.3] compute average $<E>_i$ at temperature θ
  [2.4] if $(<E>_i - <E>_{i-1})/<E>_i < \varepsilon$ then display results and stop
[1.1] end loop over temperature index i.

As noted already, an essential feature of the Metropolis procedure is the possibility to include states which increase the value of the objective function. This allows eventual escape from local minima of

E in the configuration space, thus reducing the chances of entrapment in a suboptimal solution. Current areas of active research address the development of methods for effective selection of new configurations (i.e., selection of the function F), as well as the determination of appropriate annealing schedules (i.e., selection of annealing temperatures θ_i and sample sizes at these temperatures).

IV.3. Simple Illustration for Static Load Balancing

Consider a d-dimensional hypercube multiprocessor with $N = 2^d$ identical nodes indexed by n [with n = 1,. . .N]. Consider furthermore a set T of communicating processes [indexed by i, with i = 1,. . .dim T] to be executed in minimum time on this homogeneous concurrent computation ensemble. For analogy with statistical mechanics let each process i in T correspond to a 'particle'. Let n_i define the 'position' of task (or particle), i, i.e., n_i will point to the processor on which task i is currently located.

Following Fox and Jefferson,[22] we shall identify the kinetic energy K_i of particle i with the non-message-passing portion of the execution time of the corresponding process. Our methodology, however, does not restrict the K_i's to constant values. Rather, we model each process i as a finite automaton having a set of states s_i, each state $s_{i\mu}$ with its own sequential execution time $K_i[s_{i\mu}]$ and its own set of output messages. Such states correspond to different computational paths within the process; state transition is event driven and we can include dependency on message transmission delays.

A potential energy V_i will represent the total time spent by process i for communication. Specifically, we take

$$V_i \simeq \sum_{j\neq i} \sum_{\mu} \{\psi_{ij} + H(n_i \oplus n_j)[\psi + L_{ij}^{\mu}(s_i,s_j)/b]\} = \sum_{j\neq i} U_{ij} \qquad (2)$$

where U_{ij} measures the total message-passing time between processes i and j. As shown, U depends not only on the relative positions n_i and n_j through their Hamming distance H, but also on such important factors as channel baud rate, b, and process state sets s_i and s_j; communication initialization, message routing strategy and packet queuing techniques are grouped in the 'overhead' functions ψ_{ij} and ψ, and the 'length' of message μ between processes i and j is denoted by L^{μ}_{ij}. In an actual implementation (see following section) one needs also to be concerned with eventual handling of precedence constraints. Notice that U induces an 'attractive force' between processes, i.e., the communication time will be minimized if communication processes are clustered on the same or nearby processors. The corresponding objective function is then

$$E_{KU} = \sum_i K_i[s_i] + \sum_i \sum_{j \neq i} U_{ij}/2 \qquad (3)$$

To induce processes to use efficiently all system resources we introduce load balancing constraints. We assume that each node of the homogeneous ensemble has M resource capacities (CPU, communication channels,...) denoted R_m. Let r_{im} represent the requirement of process i for resource m. The average load per node for resource m is then simply

$$\Lambda_m = \frac{1}{NR_m} \sum_i r_{im} \qquad (4)$$

The actual load of node n is

$$\lambda_{nm} = \frac{1}{R_m} \sum_i r_{im} \, \delta_{nn_i} \qquad (5)$$

where δ denotes the Kronecker symbol. The load balance constraints can then be expressed as

$$|\lambda_{nm} - \Lambda_m| \leq \varepsilon_{nm} \qquad (6)$$

for n = 1,...N and m = 1,...M, where ε_{nm} are real positive constants which characterize the system throughout requirements. In order to convert this formulation into an unconstrained optimization problem, we define the new objective function

$$E_{LB} = E_{KU} + \sum_m \sum_n w_{mn} [\lambda_{nm} - \Lambda_m]^2 \triangle_{mn} \qquad (7)$$

where the step function

$$\Delta_{mn} = \begin{cases} 0 \text{ if } |\lambda_{nm} - \Lambda_m| - \varepsilon_{nm} \leq 0 \\ 1 \text{ otherwise} \end{cases} \tag{8}$$

serves to ignore the constraint whenever the process configuration on the hypercube is within the feasible range, and to treat the constraint as an equality if the load balance requirements are violated. The positive constants ω_{mn} are weighting factors specified by the user, depending on how strongly he feels about each constraint. In our load balancing problem the constraints of Eq. (6) represent mathematical approximations to physical phenomena which are not known precisely. The value of ω_{mm} can be chosen commensurate with the modeling accuracy of the (m,n)-th constraint. In particular, when $\omega_{mn} = 0$, the constraint is ignored, and when $\omega_{mn} = \infty$, the constraint is satisfied exactly. After some algebraic manipulations, Eq. (7) can be rewritten as

$$E_{LB} = E_{KU} + \sum_i \sum_j W_{ij} \tag{9}$$

where

$$W_{ij} = \sum_m \frac{r_{im} r_{jm}}{R_m^2} \sum_n w_{mn} \left[\delta_{nn_i} \delta_{nn_j} - \frac{2\delta_{nn_i}}{N} + \frac{1}{N^2} \right] \Delta_{mn} \tag{10}$$

can be interpreted as a 'repulsive potential' which induces the processes to spread out over the hypercube's resources. For static systems the minimization is carried out using the procedure described in Section IV.2 above.

IV.4. Generalization of Simulated Annealing

In the standard simulated annealing methodology, the acceptance of a configuration is based on the Metropolis criterion which subsumes a canonical distribution, i.e., it is implicitly assumed that the number of 'particles' in the system remains constant. In terms of an optimization problem this means that the number of degrees of freedom remains constant in time. The validity of this assumption is highly questionable for many event-driven applications, particularly in the area of intelligent autonomous systems, where the process structure (i.e., the interrelationship between the various tasks in the system) may evolve dynamically, e.g., new processes are created, others are annihilated. In an SDI application, for example, warheads and decoys can be modeled as computational 'processes', the number of which can vary dramatically in time. The standard methodology also requires equilibrium to be reached at each 'temperature'. This is a lengthy process and thus a serious obstacle for applications requiring near real-time performance.

To handle problems in a real-time environment we are currently exploring the applicability of nonequilibrium statistical mechanics techniques. Specifically, if we interpret optimization as a diffusion and branching process in hypercube space, an appropriate Fokker-Planck-type equation[19] can be formulated. From a formal expansion of the Fokker-Planck function P in a complete set of eigenvectors and eigenvalues, one should be able to readily demonstrate that all 'excited states' decay exponentially fast, with a decay constant given by the excitation energy from the ground state. In other teams, our optimization problem should converge exponentially fast to the lowest 'energy state', i.e., to the global optimum. Alternatively, if we choose a 'path-integral' representation,[23] importance sampling techniques could be used to increase speed and efficiency by giving more weight to more probable states.

IV.5. Precedence Constraints

We are exploring both static and dynamic load balancing. The implementation is being carried out within the framework of the ROSES system.[14] The current version of ROSES[24] was developed to provide uniquely powerful scheduling capabilities for mapping precedence-constrained task graphs onto a concurrent computation ensemble. Although this problem is NP-complete, ROSES achieves near-optimal solutions by combining heuristic techniques for handling time complexity with special instances of abstract data structures to handle space complexity. Currently ROSES assumes a non-preemptive scheduling approach: whenever there is a processor ready to be assigned a task, an individual assignment is made. Each assignment corresponds to a 'base point', i.e., one may vary the scheduling solution only by changing each individual assignment, while the time point and processor under consideration remain unaltered. At each base point, all tasks ready to be assigned (because their precedence requisites are satisfied) constitute a 'set of alternatives' (or A-set). The A-set is constructed and updated in such a way as to continually satisfy the precedence constraints. Choosing a process (i.e., task) for execution from an A-set is guided by heuristics combined with graph-theoretic impasse detection techniques.

ROSES is being modified so that static load balancing would involve the implementation of the simulated annealing algorithm at the A-set level. The modified ROSES kernel would then be run on the NCUBE controller board, prior to task execution on the nodes.

To implement dynamic load balancing we need to significantly extend the ROSES methodology. In particular, we need to allow for task preemption, and to provide additional support by developing three classes of algorithms to be implemented by the VERTEX operating system resident on each node of the hypercube. The first class will contain information exchange algorithms, to be responsible for the continuous exchange of load information and 'task bidding' data between the processors. The second class of algorithms will be used by each processor to monitor its own load on a continuous basis in order to determine whether it can guarantee the execution of newly arriving tasks, or whether such tasks should be migrated. The third class of algorithms will handle process migration; mechanisms need to be implemented to move both code and data, and to reroute the logical communication paths. This should provide a significant measure of dynamic balancing on a short range, fast response scale. ROSES, at a higher hierarchical level, would then attempt to drive the system to global equilibrium, by applying generalized simulated annealing beyond the A-set level.

V. AN IMPLEMENTATION EXPERIMENT: INVERSE DYNAMICS EQUATIONS FOR A ROBOT MANIPULATOR

Let us now address some of the practical considerations related to the implementation of robot-related algorithms on the NCUBE concurrent computer. To fix the ideas, we consider in the following the solution of the inverse dynamics equations of a manipulator arm. This problem was chosen for illustrative purposes, i.e., because it is relatively 'simple,' while exhibiting the nonlocal communication and structural irregularity characteristics of interest.

V.1. Newton-Euler Inverse Dynamics

Several state-of-the-art formalisms are currently available to efficiently solve the inverse dynamics problem of a serial link manipulator, in which forces or torques are predicted based on desired motion. In the Newton-Euler formalism the equations for each link are written in link-fixed coordinate systems in order to simplify the calculation of the inertia tensors. A set of recurrence relations allow the angular velocities, angular accelerations and linear accelerations at the center-of-mass of each link to be successively calculated from the base to the end effector. Net forces and torques acting on each link's center-of-mass are then obtained. Forces and torques acting at the joints are subsequently calculated in a recursion from the "hand" to the base. Joint actuator torques or forces are determined from a knowledge of the orientation of each joint. The detailed derivations can be found in Luh and Lin's seminal paper.[25]

For the sake of brevity, we include here (in Table 1) only our proposed partition of the equations of motion into a set of tasks since it differs from Luh's. In the nomenclature for Table 1, as given below, all vectors are generally expressed in the i-th coordinate system:

w_i = angular velocity of link i
$\dot{w}_i$ = angular acceleration of link i
$\ddot{p}_i$ = linear acceleration of link i
F_i = net force acting on link i
r_i = net torque acting on link i about the center-of-the-mass
P_i^* = origin of the i-th coordinate system with respect to the (i − 1) th
r_i^* = position of center of mass of link i with respect to the origin of link i
J_i = inertia tensor about center of mass of link i
A_i^{i-1} = frame transformation matrix from (i − 1) -th to i-th coordinate systems (denoted as A_i^- in the table); by

analogy, A_i^+ will denote the frame transformation A_i^{i+1}
f_i = force exerted on link i by link i − 1
γ_i = moment exerted by link i − 1 on link i

Typical initial conditions are $w_o = \dot{w}_o \equiv 0$; $\ddot{p}_o = gz_o$, thereby absorbing gravity into the initial acceleration to reflect the simplified form of the force balance equation

$$f_i = F_i + A_i^+ f_{i+1} \tag{12}$$

Initial conditions for the backward recursion are obtained from the specification of f_{N+1} and γ_{N+1}, the external force and moment exerted on the hand. The joint actuator torques or forces are simply (omitting friction):

$$\tau_i = \tilde{\gamma}_i \cdot A_i^- z_{i-1} \quad \text{or} \quad \phi_i = \tilde{f}_i \cdot A_i^- z_{i-1} \tag{13}$$

Table 1. Robot Inverse Dynamics Computational Tasks Nomenclature (Newton-Euler Formalism, six link manipulator)

Task #	Rotational Link	Prismatic Link
1−6	$\mathbf{w}_i = \mathbf{A}_i^-(\mathbf{w}_{i-1} + \mathbf{z}_{i-1}\, \dot{q}_i)$	$\mathbf{w}_i = \mathbf{A}_i^- \mathbf{w}_{i-1}$
7−12	$\dot{\mathbf{w}}_i = \mathbf{A}_i^-(\dot{\mathbf{w}}_{i-1} + \mathbf{z}_{i-1}\, \ddot{q}_i + \mathbf{w}_{i-1} \times \mathbf{z}_{i-1}\dot{q}_i)$	$\dot{\mathbf{w}}_i = \mathbf{A}_i^- \dot{\mathbf{w}}_{i-1}$
13−18	$\mathbf{V}_i^{(1)} = \mathbf{w}_i \times (\mathbf{w}_i \times \mathbf{p}_i^*)$	$\mathbf{V}_i^{(1)} = \mathbf{w}_i \times (2\mathbf{A}_i^- \mathbf{z}_{i-1}\dot{q}_i + \mathbf{w}_i \times \mathbf{p}_i^*)$
19−24	$\ddot{\mathbf{p}}_i = \mathbf{A}_i^- \ddot{\mathbf{p}}_{i-1} + \dot{\mathbf{w}}_i \times \mathbf{p}_i^* + \mathbf{V}_i^{(1)}$	$\ddot{\mathbf{p}}_i = \mathbf{V}_i^{(1)} + \mathbf{A}_i^-(\ddot{\mathbf{p}}_{i-1} + \mathbf{z}_{i-1}\ddot{q}_i) + \dot{\mathbf{w}}_i \times \mathbf{p}_i^*$
25−30	$\mathbf{V}_i^{(2)} = \mathbf{w}_i \times [\mathbf{w}_i \times \mathbf{r}_i^*]$	
30−36	$\mathbf{F}_i = m_i[\mathbf{V}_i^{(2)} + \dot{\mathbf{w}}_i \times \mathbf{r}_i^* + \ddot{\mathbf{p}}_i]$	
37−42	$\Gamma_i = \mathbf{J}_i\dot{\mathbf{w}}_i + \mathbf{w}_i \times (\mathbf{J}_i\mathbf{w}_i)$	
43−48	$\mathbf{V}_i^{(3)} = \Gamma_i + (\mathbf{p}_i^* + \mathbf{r}_i^*) \times \mathbf{F}_i$	
49−54	$\mathbf{f}_i = \mathbf{F}_i + \mathbf{A}_i^+\mathbf{f}_{i+1}$	
55−60	$\mathbf{V}_i^{(4)} = \mathbf{p}_i^* \times (\mathbf{f}_i - \mathbf{F}_i)$	
60−66	$\gamma_i = \mathbf{A}_i^+\gamma_{i+1} + \mathbf{V}_i^{(3)} + \mathbf{V}_i^{(4)}$	

(*) $\dot{q}_i$ and $\ddot{q}_i$ are the first and second time derivatives of the generalized coordinates.

(**) the joint index i runs from 1 to 6 for tasks 1 to 48, and from 6 to 1 for tasks 49 to 66.

V.2. Parallel Algorithms for Inverse Dynamics

The pioneering work of Luh and Lin on scheduling of parallel computations for a computer controlled mechanical manipulator[26] has established a solid foundation for further research and development of parallel algorithms for robot dynamics. In their approach, one CPU is associated with each joint (or link). Each CPU is connected both to a primary memory, which stores local programs and data, and to a "common" memory, located between adjacent CPUs, which stores common data and information necessary for interprocessor communication. The Newton–Euler formalism provides the computational framework. Because of the dynamic coupling between adjacent links, Luh and Lin developed a parallel formulation of the inverse dynamics problem in terms of a multicomputer task-scheduling optimization problem under series-parallel precedence constraints. Their solution, based on a generalization of the branch-and-bound algorithm, exhibits, however, several significant limitations. Most importantly, the bijective computer-to-link mapping is ideally not desirable for real-world applications, since it is subject to single-point failures. Furthermore, because of the underlying topology, the system suffers from severe load unbalance, i.e., some processors are very underutilized. Finally, the issues of intertask communication and synchronization are not directly addressed.

In recent years, there has been an increased interest in the development of parallel algorithms for inverse robot dynamics.[26–30] As reported by Kasahara and Narita,[30] it seems that most of these studies do not involve an implementation on an actual multiprocessor system. Results are therefore often presented in terms of "number of additions and multiplications" and their theoretical equivalent of processor clock cycles,[29] ignoring many fundamental constraints of multiprocessing such as communication overheads or saturation effect bottlenecks. Furthermore, the emphasis is in general on architectures fully dedicated to a specific algorithm formulation. Our approach is to use a single large-scale multiprocessor system, the NCUBE, for the concurrent solution of all major algorithms involved in the operation of the autonomous robot. Since the NCUBE is reconfigurable (i.e., subcubes of specified dimension can be allocated in real time), this approach attempts to make the best possible use of the available computing resources.

V.3. The Inverse Dynamics Task Graph

One of the first, and perhaps one of the most important steps in attempting to solve a problem on a concurrent computation ensemble is its decomposition into a set of tasks (or processes). For the Newton–Euler inverse dynamics equations a tentative decomposition is given in Table–1. It involves 66 computational tasks for a 6 degrees of freedom manipulator. Problem decomposition induces precedence constraints among the tasks, and the distributed nature of the computational system translates that into message passing requirements.

Task graphs in the parallel processing literature generally use nodes to denote tasks and edges to describe communication links or precedence constraints. This convention, however, appears to us quite unnatural. Hence, a slightly different model is employed to facilitate the mapping of the inverse dynamics algorithm onto the ensemble computer. In the ROSES formalism,[24] edges denote tasks (computational or communication), while nodes represent synchronization points for the precedence constraints. The complete task graph, including all message passing requirements, is generated automatically by ROSES, for a given problem decomposition. A typical example is shown in Tables 2 and 3 for the Stanford arm. The task graph is then embedded into a concurrent computation ensemble, the exact configuration (e.g., hypercube order) of which is a user-specified input.

ROSES has currently the capability of producing two kinds of information. For rapid prototyping studies it simply generates a multiprocessor schedule. Since this option does not involve subsequent task execution on the actual hypercube, the timing estimates provided are based on typical algorithmic parameters such as the number of floating point operations per task. Then, using the appropriate processing system parameters, the expected performance of the hypercube for the particular problem decomposition can be closely simulated, without having the burden of actual code generation and debugging.

When an acceptable problem decomposition has been achieved, task coding can begin, and an alternative option in ROSES must be used. This option generates the basic control information required

Table 2 Computational Tasks for the Solution of the Newton–Euler Robot Inverse Dynamics Equations as Input to ROSES

Task ID	Mul #	Add #	Pr. C. #	Prec. Constr. IDs		
1	8	6	1	0	0	
2	8	6	1	1		
3	8	5	1	2		
4	8	6	1	3		
5	8	6	1	4		
6	8	6	1	5		
7	10	8	1	0		
8	10	8	2	1	7	
9	8	5	1	8		
10	10	8	2	3	9	
11	10	8	2	4	10	
12	10	8	2	5	11	
13	12	6	1	1		
14	12	6	1	2		
15	15	8	1	3		
16	12	6	1	4		
17	12	6	1	5		
18	12	6	1	6		
19	14	14	2	7	13	
20	14	14	3	8	14	19
21	14	15	3	9	15	20
22	14	14	3	10	16	21
23	14	14	3	11	17	22
24	14	14	3	12	18	23
25	12	6	1	1		
26	12	6	1	2		
27	12	6	1	3		
28	12	6	1	4		
29	12	6	1	5		
30	12	6	1	6		
31	9	9	3	7	19	25
32	9	9	3	8	20	26
33	9	9	3	9	21	27
34	9	9	3	10	22	28

Table 2 Computational Tasks for the Solution of the Newton–Euler Robot Inverse Dynamics Equations as Input to ROSES

Task ID	Mul #	Add #	Pr. C. #	Prec. Constr. IDs		
35	9	9	3	11	23	29
36	9	9	3	12	24	30
37	12	6	2	1	7	
38	12	6	2	2	8	
39	12	6	2	3	9	
40	12	6	2	4	10	
41	12	6	2	5	11	
42	12	6	2	6	12	
43	6	9	2	31	37	
44	6	9	2	32	38	
45	6	9	2	33	39	
46	6	9	2	34	40	
47	6	9	2	35	41	
48	6	9	2	36	42	
49	8	8	1	36		
50	8	8	2	35	49	
51	8	8	2	34	50	
52	8	8	2	33	51	
53	8	8	2	32	52	
54	8	8	2	31	53	
55	6	6	2	36	49	
56	6	6	2	35	50	
57	6	6	2	34	51	
58	6	6	2	33	52	
59	6	6	2	32	53	
60	6	6	2	31	54	
61	8	11	2	48	55	
62	8	11	3	47	56	61
63	8	11	3	46	57	62
64	8	11	3	45	58	63
65	8	11	3	44	59	64
66	8	11	3	43	60	65

Table 3 Robot Inverse Dynamics Task Graph as Generated by ROSES.

Nodes Head	Nodes Tail	Cost Mul	Cost Add	Edge #	Nodes Head	Nodes Tail	Cost Mul	Cost Add	Edge #
3	2	8	6	1	103	102	8	8	51
5	4	8	6	2	105	104	8	8	52
7	6	8	5	3	107	106	8	8	53
9	8	8	6	4	109	108	8	8	54
11	10	8	6	5	111	110	6	6	55
13	12	8	6	6	113	112	6	6	56
15	14	10	8	7	115	114	6	6	57
17	16	10	8	8	117	116	6	6	58
19	18	8	5	9	119	118	6	6	59
21	20	10	8	10	121	120	6	6	60
23	22	10	8	11	123	122	8	11	61
25	24	10	8	12	125	124	8	11	62
27	26	12	6	13	127	126	8	11	63
29	28	12	6	14	129	128	8	11	64
31	30	15	8	15	131	130	8	11	65
33	32	12	6	16	133	132	8	11	66
35	34	12	6	17	4	3	0	0	67
37	36	12	6	18	6	5	0	0	68
39	38	14	14	19	8	7	0	0	69
41	40	14	14	20	10	9	0	0	70
43	42	14	15	21	12	11	0	0	71
45	44	14	14	22	16	3	0	0	72
47	46	14	14	23	16	15	0	0	73
49	48	14	14	24	18	17	0	0	74
51	50	12	6	25	20	7	0	0	75
53	52	12	6	26	20	19	0	0	76
55	54	12	6	27	22	9	0	0	77
57	56	12	6	28	22	21	0	0	78
59	58	12	6	29	24	11	0	0	79
61	60	12	6	30	24	23	0	0	80
63	62	9	9	31	26	3	0	0	81
65	64	9	9	32	28	5	0	0	82
67	66	9	9	33	30	7	0	0	83
69	68	9	9	34	32	9	0	0	84
71	70	9	9	35	34	11	0	0	85
73	72	9	9	36	36	13	0	0	86
75	74	12	6	37	38	15	0	0	87
77	76	12	6	38	38	27	0	0	88

Table 3 Robot Inverse Dynamics Task Graph as Generated by ROSES.

Nodes Head	Nodes Tail	Cost Mul	Cost Add	Edge #	Nodes Head	Nodes Tail	Cost Mul	Cost Add	Edge #
79	78	12	6	39	40	17	0	0	89
81	80	12	6	40	40	29	0	0	90
83	82	12	6	41	40	39	0	0	91
85	84	12	6	42	42	19	0	0	92
87	86	6	9	43	42	31	0	0	93
89	88	6	9	44	42	41	0	0	94
91	90	6	9	45	44	21	0	0	95
93	92	6	9	46	44	33	0	0	96
95	94	6	9	47	44	43	0	0	97
97	96	6	9	48	46	23	0	0	98
99	98	8	8	49	46	35	0	0	99
101	100	8	8	50	46	45	0	0	100
54	7	0	0	106	48	25	0	0	101
56	9	0	0	107	48	37	0	0	102
58	11	0	0	108	48	47	0	0	103
60	13	0	0	109	50	3	0	0	104
62	15	0	0	110	52	5	0	0	105
62	39	0	0	111	108	107	0	0	162
62	51	0	0	112	110	73	0	0	163
64	17	0	0	113	110	99	0	0	164
64	41	0	0	114	112	71	0	0	165
64	53	0	0	115	112	101	0	0	166
66	19	0	0	116	114	69	0	0	167
66	43	0	0	117	114	103	0	0	168
66	55	0	0	118	116	67	0	0	169
68	21	0	0	119	116	105	0	0	170
68	45	0	0	120	118	65	0	0	171
68	57	0	0	121	118	107	0	0	172
70	23	0	0	122	120	63	0	0	173
70	47	0	0	123	120	109	0	0	174
70	59	0	0	124	122	97	0	0	175
72	25	0	0	125	122	111	0	0	176
72	49	0	0	126	124	95	0	0	177
72	61	0	0	127	124	113	0	0	178
74	3	0	0	128	124	123	0	0	179
74	15	0	0	129	126	93	0	0	180
76	5	0	0	130	126	115	0	0	181

Table 3 Robot Inverse Dynamics Task Graph as Generated by ROSES.

Nodes		Cost		Edge	Nodes		Cost		Edge
Head	Tail	Mul	Add	#	Head	Tail	Mul	Add	#
76	17	0	0	131	126	125	0	0	182
78	7	0	0	132	128	91	0	0	183
78	19	0	0	133	128	117	0	0	184
80	9	0	0	134	128	127	0	0	185
80	21	0	0	135	130	89	0	0	186
82	11	0	0	136	130	119	0	0	187
82	23	0	0	137	130	129	0	0	188
84	13	0	0	138	132	87	0	0	189
84	25	0	0	139	132	121	0	0	190
86	63	0	0	140	132	131	0	0	191
86	75	0	0	141	1	123	0	0	192
88	65	0	0	142	1	125	0	0	193
88	77	0	0	143	1	127	0	0	194
90	67	0	0	144	1	129	0	0	195
90	79	0	0	145	1	131	0	0	196
92	69	0	0	146	1	133	0	0	197
92	81	0	0	147					
94	71	0	0	148					
94	83	0	0	149					
96	73	0	0	150					
96	85	0	0	151					
98	73	0	0	152					
100	71	0	0	153					
100	99	0	0	154					
102	69	0	0	155					
102	101	0	0	156					
104	67	0	0	157					
104	103	0	0	158					
106	65	0	0	159					
106	105	0	0	160					
108	63	0	0	161					

by the multitasking programs residing on each node. In particular, ROSES provides a list of assigned tasks for each node of the ensemble. For each task this list specifies:

a. the number of messages to be received; for each message its originating node and its 'message type,'
b. the number of messages to be sent; for each message its destination node and its message type.

A message type is sometimes treated as a 'packed word,' the encoding of which will be explained below. It should be noted again that, instead of an absolute time schedule, ROSES now provides relative intertask synchronization parameters.

V.4. The GOLEM Code

The Newton-Euler inverse dynamics formalism has been implemented in a computer node named GOLEM. The emphasis was on generality, i.e., we have attempted to avoid 'tailoring' the code for a specific robot arm. In particular, there is no limit on the number of joints, their type (rotational or prismatic) or on the manipulator configuration that GOLEM can handle. All matrices, vectors and scalars involved in the solution of the inverse dynamics equations (see Table−1) are stored in two one-dimensional arrays, denoted BC and NC, for reals and integers respectively. Storage requirements are calculated at run time, pointers are defined, and memory is allocated dynamically.

In order to enhance potential interactions with other robot activities sharing the hypercube, the architecture of GOLEM is modular. Key stages in the computational flow are modular interface, dynamic memory allocation, database processing, model initialization (e.g., frame transformation matrices, inertia tensors), actual solution of the equations and display of the results. Two modes of operation are available. In the sequential mode all computations take place on the Intel 80286/80287 processors resident on the NCUBE Peripheral Subsystem. In the concurrent mode the equations are solved on the hypercube nodes.

Let us now briefly examine the basic steps involved in setting up the concurrent computation. Several system functions are provided[31] that allow use of the ensemble by FORTRAN AND C programs. All functions return an integer*4 result; negative values correspond to an error code. First one needs to allocate the requested hypercube:

$$\text{ISH} = \text{NOPEN (NOH)} \tag{14}$$

In the above statement NOH is an integer*2 variable giving the hypercube order. The channel number to be used when referencing that hypercube is the integer*2 variable ISH. The second step involves the loading of programs onto the nodes of the previously allocated hypercube. One can load either the same program onto all nodes or a different program onto each node. The only restriction in this preliminary release of the NCUBE system is the limit of one program per node. Thus, it is currently the responsibility of the user to provide multitasking capabilities at the node level, if multitasking is required by his particular solution algorithm. This is the case for GOLEM. Our approach is to download the same program to all nodes, i.e., in principle each node could execute all tasks in Table−I. However, only the tasks scheduled to be run on a particular node will be activated. To achieve this capability, we had to develop a 'focused addressing' algorithm which uses the appropriate switching and synchronization information provided by ROSES. The download statement has the form:

$$\text{istat} = \text{NLOAD (ISH, codex, IN, WA, NWA)} \tag{15}$$

where IN is an integer*2 variable giving the number of the node that is to receive the program, and codex is a character string that contains the name of the file to be loaded. The user must provide a work buffer (WA) that is at least as large as the size of the node program. The variable NWA gives the size of the working area in bytes. For large applications one must, of course, send different code segments to each node, but the focused addressing scheme is still valid.

The third step involves the sending of data to the nodes. For a particular node IN, this is accomplished using the statements:

$$\text{istat} = \text{NWRITE (ISH, BC, NLOR*4, IN, mtypr)} \quad (16)$$

$$\text{jstat} = \text{NWRITE (ISH, NC, NLOI*4, IN, mtypi)} \quad (17)$$

where the message-type parameters mtypr and mtypi simply indicate whether the data in the message are reals or integers. We download the complete working memory (arrays BC and NC of run-time word-sizes NLOR and NLOI, respectively) to all nodes. This is consistent with the approach outlined above. Furthermore, the array pointers have been stored in NC, and can be selectively retrieved at each node for maximum flexibility. Clocks are now initialized and we proceed with the solution of the equations on the hypercube ensemble.

Finally, we must receive the problem responses from the appropriate nodes. The preferred implementation is clearly via interrupts. A possible alternative is to continuously test for any arriving message from any node using the system function NTEST. The message is then retrieved with the statement

$$\text{isize} = \text{NREAD (ISH, WB, NWB, IN, LL)} \quad (18)$$

where WB is a buffer of size NWB bytes. This buffer is assumed to be large enough to hold the longest message that one would expect to receive from any node. IN denotes the source (originating node) of the message read. The message type LL is defined as the pointer to the subarray in BC (or NC) where the incoming message needs to be stored. Since a positive value of 'isize' represents the size in bytes of the received message, this is accomplished as follows, assuming again 4-byte words

$$\text{NML} = \text{isize/4}$$

$$\text{CALL VECEQV (BC(LL+1), WB, NML)} \quad (19)$$

After all expected messages have been received, timing is performed, the hypercube is deallocated, and the final results are displayed.

The communication primitives used on the nodes are semantically and syntactically quite similar to the ones described above for the host peripheral subsystem, and we will not enter into details. The most significant differences involve the use of an additional primitive, WHOAMI, and the implementation of the message-type concept. The subroutine WHOAMI is called by each node program to establish the node's identity and its relationship to the host for the current hypercube partition:

CALL WHOAMI (IDN, IDP, IDH, NOH) (20)

All arguments are integer*2 variables and denote, respectively, the identities of the node, the calling process and the host, and the hypercube order allocated to the problem. The message-type parameters used for internode communication are defined as 16-bit 'packed words.' The six most significant bits represent a pointer position (not value) in array NC. The next 9 bits encode task identity information obtained from ROSES. The least significant bit is used to determine whether the data in the message are reals or integers.

V.5. Preliminary Results

We present now the results of our inverse dynamics calculations. These results are preliminary in the sense that the CESAR NCUBE machine is a ''beta-site' prototype. The 'time score' of the solution should be expected to change in the commercial ('production') version of the NCUBE/ten due to probable changes in system performance characteristics such as processor clock rate, operational status of the instructions or compiler optimization capabilities. Thus, rather than list absolute timings for the computations, we prefer to show here more conservative trends such as expected levels of processor load balance (throughput). The assignment of the tasks in Table-1 to nodes of a two-dimensional hypercube is given in Table-4 for the forward and backward recursions. The processor utilization results shown in Table-5 refer to the most unbalanced node and are given for two, three and four processors. The improvement in performance observed (a factor of two in the non-asymptotic domain over previously reported results[26]) is

Table 4 Task Assignment of the Inverse Dynamics Equations for a 2-Dimensional Hypercube

Node 0	Node 1	Node 2	Node 3
1	7	2	25
13	8	14	3
19	9	4	15
20	10	16	11
21	5	28	6
22	17	12	18
23	29	30	41
24	42	27	26
36	40	34	39
49	33	46	45
55	32	35	38
61	31	47	44
62	37	48	
63	43	50	
64	51	56	
65	57	52	
66	53	58	
	59	54	
		60	

Table 5 Processor Load Balance for the Solution of the Inverse Dynamics Equations of the Stanford Manipulator

Item		This Work			Luh & Lin[26]
Processors	1	2	3^+	4^*	6
Load**	1	.96	.95	.61	.37
Speed-up	1	1.96	2.87	3.22	2.56

$^+$We allocate an order-2 cube, but schedule tasks on three nodes only.

*Assymptotic domain critical length reached: using more than 4 processors for this simple example wastes resources.

**Load factor for most unbalanced node in system.

Table 6 Benchmark Parameters for the Solution of the Newton Euler Equations of a Robot Manipulator*

Task ID	Mul #	Add #	Pr. C. #	Prec. Constr. IDs			Task ID	Mul #	Add #	Pr. C. #	Prec. Constr. IDs		
1	0	0	1	0			47	3	6	3	44	45	46
2	0	0	1	0			48	3	0	1	43		
3	2	0	1	1			49	9	3	1	41		
4	1	0	1	1			50	0	3	2	48	49	
5	3	0	3	2	3	4	51	4	3	1	41		
6	1	0	1	1			52	2	0	1	41		
7	2	0	1	1			53	4	5	2	43	52	
8	1	0	1	1			54	4	2	1	44		
9	4	2	2	1	8		55	6	2	1	51		
10	2	0	1	2			56	2	0	1	53		
11	6	2	1	7			57	0	6	3	54	55	56
12	2	0	1	9			58	12	6	1	51		
13	0	6	3	10	11	12	59	6	3	1	53		
14	10	4	1	7			60	3	6	3	57	58	59
15	4	1	1	9			61	9	6	1	53		
16	3	6	3	13	14	15	62	15	9	1	51		
17	3	0	1	9			63	0	3	2	61	62	
18	9	3	1	7			64	6	3	1	60		
19	0	3	2	17	18		65	0	3	2	63	64	
20	0	0	2	7	9		66	4	5	1	47		
21	2	0	1	20			67	4	2	1	65		
22	9	2	1	20			68	4	1	1	47		

Table 6 Benchmark Parameters for the Solution of the Newton Euler Equations of a Robot Manipulator*

Task ID	Mul #	Add #	Pr. C. #	Prec. Constr. IDs			Task ID	Mul #	Add #	Pr. C. #	Prec. Constr. IDs			
23	0	1	1	13			69	0	6	3	50	67	68	
24	0	6	3	21	22	23	70	4	5	2	37	66		
25	6	2	1	20			71	4	2	1	69			
26	2	0	1	20			72	4	1	1	37			
27	3	6	3	24	25	26	73	0	6	3	40	71	72	
28	3	0	1	20			74	4	5	2	27	70		
29	9	3	1	20			75	6	2	1	70			
30	0	3	2	28	29		76	4	4	2	73	75		
31	4	3	1	20			77	2	0	1	27			
32	2	0	1	20			78	0	6	4	30	74	76	77
33	4	5	2	20	32		79	0	3	2	16	74		
34	4	2	1	24			80	2	0	1	74			
35	10	4	1	31			81	0	2	2	78	80		
36	4	1	1	33			82	4	1	1	16			
37	3	6	3	34	35	36	83	0	6	3	19	81	82	
38	3	0	1	33			84	4	5	2	5	79		
39	9	3	1	31			85	4	1	1	79			
40	0	3	2	38	39		86	4	4	2	83	85		
41	4	3	1	31			87	4	1	1	5			
42	2	0	1	31			88	0	6	3	6	86	87	
43	4	5	2	33	42									
44	4	2	1	34										
45	10	4	1	41										
46	4	1	1	43										

(*)Adapted from Tables VII and VIII of Ref. 26, where the equations were 'specialized' to the Stanford 6-DOF manipulator

Table 7 ROSES Benchmark Results. Solution of the Newton–Euler Inverse Dynamics Equations for the Stanford 6–DOF Manipulator using the Parameters of Table VI

Number of Processos	Luh & Lin[26] ms	Kasahara & Narita[30] ms	ROSES (this work) ms	Optimal Schedule ms
1	24.80	24.83	24.80	24.80
2	—	12.42	12.43	12.40
3	—	8.43	8.49	8.43
4	—	6.59	6.67	6.38
5*	—	5.86	5.88	5.67
6	9.70	5.73	5.78	5.67
7	—	5.69	5.67	5.67
8	—	—	5.67	5.67

(*)Beginning of asymptotic domain: optimal schedule on any processor would equal the critical path of the graph.

particularly significant in light of the fact that the individual nodes themselves are much more powerful (by at least an order of magnitude), which the load balance results do not explicitly reflect. This should open the possibility for real-time control of flexible manipulators where many more degrees of freedom are involved.

The fundamental role of the task scheduler in achieving good concurrent computation efficiencies for irregular robotics problems of the type discussed above can not be overemphasized. Thus, it is important to provide well-defined benchmarks against which the performance of available codes can be tested. As a first step in that direction, we have carried out a comparison between the recent results of Kasahara and Narita,[30] the original work of Luh and Lin[26] and ROSES. To provide a fair basis for the comparison, we now assume that each processor has the same performance parameters as the ones used by Luh and Kasahara (40 μs for a floating point add, 50 μs for a floating point multiply, rather than the 2μs associated with the NCUBE processor design). Furthermore, instead of using the task partition of Table–1 and the cost estimates for the general form of the equations, we adopt the task partition and 'specialized' costs given by Luh and also used by Kasahara. For completeness their task data are reproduced here in Table-6. The results of the comparison are summarized in Table-7. The agreement between our results and those of Kasahara and Narita is excellent. However, whereas their method requires mainframe computing power,[30] ROSES runs on an Intel 80286 (IBM PC-AT, or NCUBE Peripheral Subsystem). Furthermore, notwithstanding the fact that the current version of our code is an unoptimized prototype, the time required to schedule over 200 tasks (88 computational and 140 message-passing) was approximately 12s on a 6MHz IBM PC-AT.

VI. CONCURRENT COMPUTATION, MACHINE INTELLIGENCE AND ROBOTICS

Potential implications of the recent advances in concurrent computation for the domains of machine intelligence and robotics are briefly examined in this concluding section. We limit our remarks to the areas of machine vision and manipulator dynamics, in which significant CESAR research efforts are underway.

VI.1. Machine Vision

The field of machine perception (e.g., vision, tactile sensing, etc.) has been, in particular over the last decade, closely associated with progress in computer architectures, as evidenced by the detailed bibliographies found in the literature.[32–34] One of the main objectives of this association was to help overcome severe drawbacks in artificial vision systems such as latency and propensity to fail except in simple structured domains.

In the past, the computational emphasis has essentially been on SIMD architectures for simple preprocessing tasks such as convolution and mask evaluation; pipelined architectures have been used for executing sequences such as blurring/differentiation/zero-crossing. To develop a realistic machine vision system[35] one needs to go beyond these retina-level tasks and include some of the higher level components of human perception. For example, humans seem to perceive effortlessly characteristics such as colinearity, direction, periodicity, coarseness and continuity. This suggests that a machine vision system ought to include efficient mathematical transformations that can handle these effects.

For such complex processing options, where the type of processing is influenced by the context of the image, MIMD concurrent computers are required. Despite the fact that the processing speed of neurons is slow by silicon chip standards, human perception is extremely fast. This implies that human perception achieves its speed through massive parallelism, involving perhaps billions of processing elements. The same will almost certainly be true of successful machine vision systems. To achieve this goal, the CESAR research program[35] on Human Analog Vision is investigating, from the outset, algorithms that can be executed on concurrent computer architectures. The NCUBE/ten appears to be the most advanced large-scale system presently available for this purpose.

VI.2. Robot Dynamics

For many of the applications in which mechanical manipulators are used today, real-time performance can be achieved in a well structured environment, using quasi standard bus-based multi-microprocessor architectures. Such manipulators generally consist of a base-anchored open-link articulated chain, composed of a few (typically six) rigid links connected by rotational or prismatic joints. The dynamic behavior of a manipulator is modeled, as shown in Section V, by a set of coupled, highly nonlinear equations of motion. The efficient solution of these equations is required, both for the design of advanced real-time control algorithms, as well as for carrying out 'planning' activities, either at the machine intelligence level of the autonomous robot, or at the man-machine interface for telerobotic applications.

A major emphasis in the design of future robots will be structural flexibility[36] and joint compliance.[37] Such models will increase very significantly the complexity of the equations of motion and associated control algorithms. Furthermore, since such calculations need to be carried out in a common computational framework with other robotic activities including vision, sensor fusion, navigation, . . . it is essential that the computers on board be able to operate in a concurrent fashion. We reported, in this chapter, on our preliminary experience and progress in this fundamental area.

VI.3. Conclusions

The development of concurrent computers raises several challenging issues. How powerful should each processor be? How should the processors communicate with each other? How should the workload be divided among the processors? How does one make sure that processors are not sitting idle waiting for input from other processors? To address the fundamental computational problems underlying the development of machine intelligence and robotics, we can now start from one of the most promising advances in the field of computer science, i.e., VLSI-driven hypercube architectures for concurrent computation. In the near term we plan to focus our primary efforts toward the development of an advanced hypercube operating system based on dynamic load balancing and suitable for a wide range of hard real-time applications.

ACKNOWLEDGEMENTS

Stephen Colley and John Palmer are the principal architects and designers of the NCUBE system. William Richardson led the design of the revolutionary processor which will power the entire NCUBE family of concurrent computers. Research funding for CESAR is provided by the U.S. Department of Energy, by the U.S. Air Force Wright Aeronautical Laboratories and by the U.S. Army Human Engineering Laboratory. The support of A. Zucker, F. C. Maienschein and C. R. Weisbin is greatly appreciated. Special thanks to S. M. Babcock, J. R. Einstein, M. C. G. Hall, E. Halbert, D. Jefferson, and G. deSaussure for enlightening discussions. This manuscript was expertly prepared by Ruth Lawson and Lucy Whitman.

REFERENCES

1. C. Mead and L. Conway, *Introduction to VLSI Systems*, Addison Wesley, Reading, Massachusetts (1980).
2. C.L. Seitz, *Concurrent VLSI Architectures*, *IEEE Trans. Comp., C33*, 1247 (1984).
3. G.C. Fox and S. W. Otto, *Algorithms for Concurrent Processors*, *Physics Today, 37*, #5, 50 (1984).
4. C. Lutz et al., *Design of the Mosaic Element*, 5093:TR:83, California Institute of Technology (1983).
5. Inmos Corporation, *The IMS T414 Transputer*, Colorado Springs, CO (1985).
6. J.F. Palmer, *Hypercubes: Architecture and Algorithms*, Second SIAM Conference on Parallel Processing for Scientific Computing, Norfolk, VA (November 1985).
7. J. Barhen and J.F. Palmer, *The Hypercube in Robotics and Machine Intelligence*, *Comp. Mech. Eng., CIME 5*, #4, 30 (1986).
8. J.A. Adam and M.A. Fitschetti, eds., *Star Wars: SDI, The Grand Experiment*, *IEEE Spectrum*, *22*, #9, 34 (1985).
9. J.A. Adam and P. Wallich, eds., *Star Wars: 1. Mind Boggling Complexity*, *IEEE Spectrum*, *22*, #9, 36 (1985).
10. J. Edler, A. Gottlieb and J. Lipkis, *Operating System Considerations for Large Scale MIMD Machines, Procs., 1985 International Computers in Engineering Conference, vol.* **3**, pp. 199–208, Boston, MA (August 1985).
11. C.L. Seitz, *The Cosmic Cube*, *CACM*, *28*, #1, 22 (1985).
12. P.B. Schneck et al., *Parallel Processor Programs in the Federal Government*, *IEEE Computer*, *18*, #6, 43 (1985).
13. J.K. Lenstra and A.H.G.R. Kan, *Complexity of Scheduling under Precedence Constraints*, *Oper. Res.*, *26*, 22 (1978).
14. J. Barhen, *Robot Inverse Dynamics on a Concurrent Computation Ensemble*, Proc., 1985 International Computers in Engineering Conference, vol. **3**, pp. 415–429, Boston, MA (August 1985).
15. Y.C. Chow and W.H. Kohler, *Models for Dynamic Load Balancing in a Heterogenous Multiple Processor System*, *IEEE Trans. Comp.*, *C28*, 354 (1979).
16. E.G. Coffman, *Computer and Job Shop Scheduling Theory*, J. Wiley, New York (1976).
17. R.L. Graham, E.L. Lawler, J.K. Lenstra and A.H.G.R. Kan, *Optimization and Approximation in Deterministic Sequencing and Scheduling: A Survey*, *Annals Discrete Math.*, *5*, 169 (1979).
18. M.R. Garey and D.S. Johnson, *Computers and Intractability: A Guide to the Theory of NP-Completeness*, Freeman, San Francisco (1979).
19. F. Reif, *Fundamentals of Statistical and Thermal Physics*, McGraw-Hill, New York (1965).
20. S. Kirkpatrick, C. Gelatt and M. Vecchi, *Optimization by Simulated Annealing*, *Science 220*, 671 (1983).
21. N. Metropolis, et al., *Equation of State Calculation by Fast Computing Machines*, *J. Chem. Phys.*, *21*, 1087 (1953).
22. G. Fox and D. Jefferson, informal presentations, First Conference on Hypercube Multiprocessors, Knoxville, TN (August 26–27, 1985).

23. L.S. Schulman, *Techniques and Applications of Path Integration*, J. Wiley, New York (1981).
24. J. Barhen, P.C. Chen and E. Halbert, *ROSES: A Robot Operating System Expert Scheduler*, ORNL/TM-9987, CESAR-86/09, Oak Ridge National Laboratory (October 1986).
25. J.Y.S. Luh, M.W. Walker and R.P. Paul, *On-Line Computational Scheme for Mechanical Manipulators*, *J. Dyn. Syst. Meas. Contr.*, *102*, 69 (1980).
26. J.Y.S. Luh and C.S. Lin, *Scheduling of Parallel Computation for a Computer Controlled Mechanical Manipulator*, *IEEE Trans. Syst. Man. Cyb.*, *SME-12*, 214 (1982).
27. P. Gupta, *Multiprocessing Improves Robotics Accuracy and Control*, *Computer Design*, *21*, 169 (November 1982).
28. R.H. Lathrop, *Parallelism in Manipulator Dynamics*, *Int. Jour. Rob. Res.*, *4*, #2, 80 (1985).
29. R. Nigam and C.S.G. Lee, *A Multiprocessor-Based Controller for the Control of Mechanical Manipulators*, *IEEE Jour. Rob. & Aut.*, *RA-1*, #4, 173 (1985).
30. H. Kasahara and S. Narita, *Parallel Processing of a Robot Arm Control Computation on a Multiprocessor System*, *IEEE J. Rob. Aut.*, *RA-1*, #2, 104 (1985).
31. Caine, Farber and Gordon, Inc., *Fortran 77 Guide For Use on NCUBE AXIS Systems*, Pasadena, CA (December 1985)
32. K.S. Fu and T. Ichikawa, Eds., *Special Computer Architectures for Pattern Processing*, CRC Press, Boca Raton, FL (1982).
33. K. Preston and L. Uhr, *Multicomputers and Image Processing*, Academic Press, New-York (1982).
34. M.H. Raibert and J.E. Tanner, *Design and Implementation of a VLSI Tactile Sensing Computer*, *Int. Jour. Rob. Res. 1*, #3, 3 (1982).
35. M.C.G. Hall, *The Human Analog Vision System*, personal communication (December 1985).
36. W.J. Book, *Recursive Lagrangian Dynamics of Flexible Manipulator Arms*, *Int. Jour. Rob. Res.*, *3*, #3, 87 (1984).
37. M.G. Forrest, S.M. Babcock et al., *Control of a Single Link, Two-Degree-of-Freedom Manipulator with Joint Compliance and Actuator Dynamics*, Procs., 1985 International Computers in Engineering Conference, vol. **1**, pp. 189–197 (August 1985).

INDEX